Science Education and
Pedagogy in South Africa

OMPLICATED

A Book Series of Curriculum Studies

William F. Pinar
General Editor

Volume 51

The Complicated Conversation series is part of the Peter Lang Education list.
Every volume is peer reviewed and meets
the highest quality standards for content and production.

PETER LANG
New York • Bern • Berlin
Brussels • Vienna • Oxford • Warsaw

Oscar Koopman

Science Education and Pedagogy in South Africa

PETER LANG
New York • Bern • Berlin
Brussels • Vienna • Oxford • Warsaw

Library of Congress Cataloging-in-Publication Data

Names: Koopman, Oscar, author.
Title: Science education and pedagogy in South Africa / Oscar Koopman.
Description: New York: Peter Lang, 2018.
Series: Complicated conversation: a book series of
curriculum studies; vol. 51 | ISSN 1534-2816
Includes bibliographical references and index.
Identifiers: LCCN 2017031529 | ISBN 978-1-4331-4808-8 (hardback: alk. paper)
ISBN 978-1-4331-4804-0 (pbk.: alk. paper) | ISBN 978-1-4331-4809-5 (ebook pdf)
ISBN 978-1-4331-4810-1 (epub) | ISBN 978-1-4331-4811-8 (mobi)
Subjects: LCSH: Science—Study and teaching—South Africa.
Science teachers—Training of—South Africa.
Classification: LCC Q183.4.S49 K66 | DDC 507.1/068—dc23
LC record available at https://lccn.loc.gov/2017031529
DOI 10.3726/b11660

Bibliographic information published by **Die Deutsche Nationalbibliothek.**
Die Deutsche Nationalbibliothek lists this publication in the "Deutsche
Nationalbibliografie"; detailed bibliographic data are available
on the Internet at http://dnb.d-nb.de/.

To Albert Harold who mentored me how to write

Table of Contents

Figures

Tables

Foreword

When a person attempts to unlock a door with the wrong key or tries to open a safe with the wrong code, the result can be much unnecessary frustration, anxiety and stress. The danger is that all of this emotional pressure can become so intense that a person can resort to violence in trying to force the lock of the door or the safe open. As I was reading this book I realised that many science teachers in South Africa metaphorically find themselves in a similar position as they try to unlock the minds of their learners using a particular pedagogy to see the world in a new light or to change their perceptions of the world.

This book *Science Education and Pedagogy in South Africa* as scholarly text is a contemporary, timeous and most welcome contribution that entices science teachers to 're-examine' or 'rethink' the pedagogical strategies they use in their teaching. It offers a fresh approach and represents a bold attempt by the author to explore and tackle the worrisome problems of how to plan, implement and present excellent and quality learning opportunities as a driver to enhance, promote and accelerate learner performance in science education. These imperatives to achieve quality education for all is weighed against the Department of Higher Education and Training's flagship policy intervention strategy, *Revised Minimum Requirements for Teacher Education Qualifications* (2015) and the National Department of Basic Education's (2015) visionary goals as stated in *Action Plan to 2019: Towards the Realisation of Schooling 2030.* This book is a well-crafted text, and is written in concise and persuasive language to convey the principles of teaching science. The author of this seminal publication paves the way for researchers, postgraduate science education scholars as well as pre-service and in-service science teachers to become empowered in their professional growth and development in search of appropriate pedagogies for a multicultural classroom.

During the many years that I have taught curriculum studies and have been involved with research and mentorship, I have always felt that there is a need for a reader-friendly academic text that could help me assist my students more effectively. *Science education and Pedagogy in South Africa* has filled this gap and offers insightful and thought-provoking suggestions on how to teach science effectively. It brings together and aligns *Science education and Pedagogy* firmly within the boundaries of the Curriculum and Assessment Policy Statement for a South African context. To fully understand and appreciate the main thrust of this book, I suggest that the reader consider Chapter 2 specifically as a follow-up to Chapter 1 in order to understand the challenges facing effective science teaching in South Africa. In these chapters the author eloquently and thought-provokingly elucidates the issue of whether the Physical Science curriculum promotes a 'canonical' or 'humanistic' approach to science. Moreover, these two constructs encapsulate what follows and therefore sets the tone of the book. Here, the author lays the foundation that 'canonical science as an approach' is 'exclusively factual, rigid, socially sterile and dogmatic, with a focus on abstract concepts, theories and laws in science, while "humanistic science" embraces human subjectivity, personal knowledge and lived experience juxtaposed with the broader needs of societies as a way to promote meaningful learning within a context'.

The author skilfully draws on scholarly works such as Merleau-Ponty's (1962) lived body theory, Aikenhead's (1996) border-crossing theory and Jegede's (1997) collateral learning theory, and links them with appropriate teaching strategies for science teachers to be used to critically engage learners in both Western and indigenous knowledge systems and their relevance in science education. Chapter 3 draws on the theories of Husserl, Heidegger and Foucault to investigate the relationship between knowledge and power in science education by highlighting the risks of poor science teaching. Chapter 4 could assist researchers to evolve suitable didactic practices to preserve the unity of Chemistry at all three levels of representation, while Chapter 5 proposes argumentation as a teaching method to facilitate learning in the science education classroom. In Chapter 6 the author advocates for the development of a phenomenological pedagogy through a professional development programme. In the final chapter he argues compellingly for a paradigm shift towards a science education that promotes a humanising philosophy in order to overcome the powerful web of neoliberal governmentality. This chapter is well written and I found myself agreeing with the general gist of the politics behind it. The author engages with philosophy and brings some interesting ideas to the conversation about how we can decolonise school science and university curricula in South Africa.

While the book does not claim to exhaust the field of science education, it goes a long way in providing a path along this route by focusing on phenomenological principles that could become the engine to a decolonised and Africanised way of teaching of science in South African schools and institutions of higher learning. Reading this book was a real joy, because it comprehensively captures and engages with some of the really challenging issues that plague teaching and learning in our country. This book meets a critical need and does so with distinction. I foresee its widespread use and, consequently, its strong contribution to the overall quality of teacher education, in particular Science teaching research, in South Africa and on the African continent.

Professor Michael van Wyk
05 June 2017
National Research Foundation Rated Researcher
Department of Curriculum and Instructional Studies
College of Education
University of South Africa

References

Aikenhead, G. (1996). Science education: Border crossing into the sub-culture of science. *Studies in Science Education, 27,* 1–57.

Department of Basic Education. (2015). *Action plan to 2019: Towards realisation of schooling 2030.* Pretoria: Author.

Department of Higher Education and Training. (2015). *Revised minimum requirements for teacher qualifications.* Pretoria: Author.

Jegede, O. J. (1997). Traditional cosmology and collateral learning in non-Western science classrooms. In *Report of an International Scientific Research Programme (Joint Research),* funded by the Grant-in-Aid for Scientific Research, Ibaraki, Japan.

Merleau-Ponty, M. (1962). *Phenomenology of perception* (C. Smith, Trans.). London: Routledge.

Preface

This book in many ways resonates with a number of challenges that continue to plague our education system and raises serious questions about any conception of quality teaching and learning. Although so many issues—such as overcrowded classrooms, inadequate resources, learner underachievement and minimal parental support—continue to distract teachers, it is the lack of an effective pedagogy that has the most serious impact under the present circumstances. Alexander (2008) avers that it is teacher pedagogy, rather than merely the curriculum or school organisation, that lies at the heart of effective teaching and learning. As a lecturer in Chemistry in the Faculty of Education at the Cape Peninsula University of Technology, I am reminded of thought-provoking questions a student asked in an informal conversation: How do you give equal attention to the needs of all learners, coming as they do from different races, ethnicities and backgrounds? How do you relate to learners whose language you can't even understand? It is significant that this student had a realistic attitude towards an understanding of pedagogy by searching for answers to difficult questions rather than abdicating his duty, as so many teachers in South Africa do today. It is questions like these that sometimes make me wonder whether I am preparing my students adequately for a career in teaching. It was this unsettling conviction that I am perhaps not doing justice to the courses I teach that motivated me to write this book.

What is needed today are science teachers who are not only effective in their pedagogical practices, but who understand the basic needs, cultures and lived world experiences of modern-day learners. Although this book is written for researchers, one of its aims is to provide signposts, so to speak, for science teacher educators in search of appropriate pedagogies for a multicultural classroom. From this perspective the aims of this book can be summarised in four short questions: (i) What knowledge? (ii) Whose knowledge? (iii) At whose expense? and (iv) For whose benefit? Although there are many theories and philosophical approaches that address these questions, this book investigates them mainly from the standpoint of lived experience, undergirded by a philosophy based on phenomenological principles.

Consequently, this book has three main objectives: (i) to critically analyse the Physical Science curriculum, (ii) to understand how the curriculum (re)configures the teacher's role as bearer of the curriculum and learner as recipient of the curriculum and (iii) to offer suggestions on how to teach science effectively. Philosophically, the predominant focus of this book is to investigate the relationship between a teacher's actions and its consequences. Asserting that a teacher's (pedagogical) actions have consequences means (i) that a teacher's choice of action matters more than his or her other intentions, and (ii) that implicit in a teacher's actions lies a response to the challenging yet dominant ideologies and discourses that continue to dictate the teaching and learning of Physical Science.

The introductory chapter of the book asks the following questions: (i) Does CAPS for Physical Science promote 'canonical' or 'humanistic' approach to science teaching? (ii) How does this approach to science and its associated knowledge promoted in CAPS (re)configure the pedagogical practices of teachers? In this chapter the term 'canonical science' refers to an approach that is exclusively factual, rigid, socially sterile and dogmatic, with a focus on abstract concepts, theories and laws in science, while 'humanistic science' embraces human subjectivity, personal knowledge and lived experience juxtaposed with the broader needs of societies as a way to promote meaningful learning within context.

Chapter 2 asks the following question: What teaching strategies can science teachers use to critically engage their learners and give due recognition to both Western and indigenous knowledge systems in order to resolve tensions between these two competing/complementary bodies of knowledge? Drawing on Merleau-Ponty's lived body theory, Aikenhead's border-crossing theory and Jegede's collateral learning theory, this chapter offers suggestions on how science teachers can overcome the pedagogical challenges facing them in post-apartheid South Africa.

Chapter 3 underscores the relationship between knowledge and power. It is an attempt to construct a conceptual framework to understand the continuous struggle that teachers face in an attempt to repudiate old ideas, habits of mind and obscure perceptions of science teaching that derive from their early training. Theoretically, the chapter draws on the work of Husserl, Heidegger and Foucault.

Chapter 4 investigates how science teachers can engage with the macroscopic, sub-microscopic and symbolic levels of representation in their Chemistry teaching. This chapter could assist researchers to evolve suitable didactic practices to preserve the unity of Chemistry at all three levels of representation.

Chapter 5 chronicles how insights from the process of argumentation can be used to facilitate learning in the science classroom by moving away from traditional didacticism towards argumentation as a teaching method. This chapter also offers teachers and learners an opportunity to understand the epistemic nature of their own learning about scientific phenomena through argumentation. Also included in the chapter is a detailed account of the author's own learning and teaching experiences that helped him to move away from traditional patterns of teaching science towards using dialogical argumentation as a teaching method.

Chapter 6 advocates for the development of a phenomenological pedagogy, through a professional development programme, with a focus on *knowledge by acquaintance* as a shift away from the concentration on propositional knowledge that is customary of traditional pedagogies.

The final chapter argues for a shift towards a humanising philosophy so that the powerful web of neoliberal governmentality can be broken in order for learners to be able to take up their rightful places at the centre of the educational agenda. In this concluding chapter I hope to advance the idea that lived experience as the focus of a phenomenological approach could be the most appropriate response to a decolonised teaching and learning environment in South Africa.

Oscar Koopman
27 May 2017

Acknowledgements

The author would like to thank Prof. Edwin Hees for his outstanding editorial work and Prof. Michael van Wyk for writing the Foreword. Also, a word of thanks to the *African Journal of Mathematics, Science and Technology Education* and its editor, Prof. Fred Lubben, for granting me permission to publish a revised version of my article as Chapter 4. To my friend, Paul Iwuanyanwu, thank you for your contribution in this book and your continued prayers and encouragement. This book is the result of prayers answered. I want to thank my wife, Karen, and children, Taffi and Phoebe, for their unwavering love and support. Finally, I am indebted to the Cape Peninsula University of Technology and the National Research Foundation who have provided funds for the preparation of this manuscript.

Acknowledgments

Abbreviations

ANC	African National Congress
C2005	Curriculum 2005
CAPS	Curriculum and Assessment Policy Statement
CASS	Continuous Assessment
CAT	Contiguity Argumentation Theory
CNE	Christian National Education
CNS	Central Nervous System
DAIM	Dialogical Argumentation Instructional Model
DBE	Department of Basic Education
DET	Department of Education and Training
DOE	Department of Education
EK	Embodied knowledge
FET	Further Education and Training
FP	Fundamental pedagogics
FRD	Foundation for Research and Development
GET	General Education and Training
IK	Indigenous knowledge
IKS	Indigenous Knowledge Systems
NATED	National Assembly Training and Education Department
NCS	National Curriculum Statement

NSC	National Senior Certificate
OBE	Outcomes-Based Education
PDP	Professional Development Program
POP	Phenomenological Observation Protocol
RNCS	Revised National Curriculum Statement
SA	Specific Aims
SAG	Subject Assessment Guidelines
TAP	Toulmin's Argumentation Pattern
UNICEF	United Children's Emergency Fund
WS	Western Science

CHAPTER ONE

Does Caps for Physical Science Promote Canonical or Humanistic Science?

Introduction

In South Africa, and many other nations across the globe, school science is governed and shaped by policy-makers whose main focus is on preparing learners for 'an increasingly technological, urbanized, competitive, global economy' (Chinn, 2007, p. 1247). Spurred on by such global goals and agendas, one of the main challenges facing school science today, in post-apartheid South Africa, is the failure of policy-makers to design and develop a curriculum that could dismantle this strongly Eurocentric epistemology that continues to dominate the curriculum. To this purpose, it is argued here that a school science curriculum should focus more on contextual humanistic science in which more, and not less, attention is afforded to human subjectivity, personal lived world experiences of learners juxtaposed with the needs of the local community. Such a curriculum provides learners with opportunities for meaningful, experiential, inquiry and place-based learning that is fundamental to scientific and environmental literacy (*Ibid.*, 2007). This is because the work of many social constructivists and phenomenologists over the years has illustrated that knowledge is: (i) not a passive commodity that flows from the mind and notes of the teacher to that of the learner, (ii) learners should not be coerced to memorise knowledge, (iii) it is a product of lived experience, (iv) knowledge is an embodied phenomena and (v) indigenous knowledge of locals should form the basis of the new knowledge to be taught. Indeed, this calls for a shift away from canonical Eurocentric science to a more culture-sensitive science curriculum that resonates with the heart and soul of learners when they are being taught.

Le Grange (2016) highlighted this shortcoming in the design and development of all post-apartheid curricula. He contends that in the design of all post-apartheid school science curricula in South Africa, curriculum planners and policymakers placed the learners and their needs on the periphery by ignoring personal knowledge, which culminated in the national curriculum frameworks that drive all post-apartheid curricula, such as Curriculum 2005 (C2005), the Revised National Curriculum Statement (RNCS) and the National Curriculum Statement (NCS). He contends that by doing so, these curricula had more in common with the apartheid curriculum framework inspired by Frank Taylor (1911) and Ralph Tyler (1949), and were merely lighter or heavier versions of the so-called 'Tylerian factory' model of schooling in which learners were synonymous with the production line of an industrial system that aims to produce specified products. He declares that the principles according to which these curricula were designed intended learners to remain consumers of predesigned knowledge. In other words, these curricula did not break with the past and have remained unchanged despite heavy criticism of their rigid, mechanistic and instrumental nature by deliberative curriculum scholars, reconceptualists and complexity theorists. Beets (2012) points out that these curricula were never aimed at producing adaptable, thinking, autonomous people, who are self-regulated and capable of cooperating with others (p. 69), but rather the continuation of the apartheid agenda for curricula which was to prepare learners to participate in the labour market (NCS, 2008). Pinar (2015) echoes that a new curriculum in its design and development features that does not value and embrace human subjectivity ('self-knowledge') runs the risk of repeating the past. He avers that a *curriculum* that is stripped of human subjectivity and historicity becomes a projection that only reproduces the past (p. 188).

Consequently, the aim of this study is to provide insight into the following two research questions: (i) Does CAPS for Physical Science promote 'canonical' or 'humanistic' science? (ii) How do this approach to science and its associated knowledge promoted in CAPS (re)configure the pedagogical practices of teachers? Canonical science is factual, rigid, socially sterile and dogmatic with a focus on concepts, theories and laws in science while humanistic science embraces human subjectivity, 'self-knowledge' and lived experience juxtaposed to the broader needs in societies as a way to promote meaningful learning within context. According to Berman and McLaughlin (1976) 'The bridge between a promising idea and its impact on students is implementation...' (p. 349). This study therefore has three objectives, that is: (i) whether CAPS promotes canonical or humanistic science,

(ii) whether these approaches to science stipulated in the CAPS can emancipate or constrain learners' understanding of science and (iii) what does this mean for effective science teaching.

Brief Historical Overview of the Science Curriculum

The word 'curriculum' has its origin in Latin for 'a course, or path, of life (curriculum vitae)' (Doll, 2008, p. 190). This term was first invoked by the French theologian and pastor of the Protestant church, Jean Calvin, in the 1500s (*ibid.*). Since the introduction of this word in the late 1500s 'curriculum' was always associated with Protestant culture, which had a bourgeois or capitalist ideology. Ramus, a headmaster of a school in the late 1500s, was the first person to apply the term in the schooling context. His idea behind curriculum was very simple by 'textbookising' knowledge, a notion for which he was vehemently critiqued by many university scholars as being 'juvenile' and 'textbookish' which was only fit for youngsters in their early teens. Hamilton (2003) points out that Ramus' instructional methods have provided the underlying paradigms in the history of modern schooling.

In the 1900s the term 'curriculum' evolved as a grand narrative (both in universities and schools) that encompasses an array of philosophical views that served as foundations of how knowledge should be methodically arranged, structured and delivered. The ideas of scholars such as John Dewey (1970), Franklin Bobbitt (1918), Frank Taylor (1911) and Ralph Tyler (1949) permeated traditional curricula. These scholars conceptualised the technical-rational or factory model, which meant that curricula were designed for neoliberal and capitalist agendas to create an effective scientific society in America. The main problem with these curricula is that they were too rigid, dogmatic, technical and theoretical in nature and did not connect with the lived world experiences of learners (Pinar, 2004).

According to Bladès (2008), since World War II the goal of school science was to create hope for a better world by engaging learners in the study of the knowledge and methodology of science. This goal was reflected in a curriculum that focused on science as a discipline of inquiry in which learners were exposed to and trained to implement the actual practice of science as a knowledge field. However, the turning point came when the Russians launched Sputnik, in 1957, signalling their intercontinental ballistic missile capabilities. This revolutionised the development of school curricula in the 1960s in Britain and the United States. To produce more scientists, engineers and mathematicians, the design of school science curricula was no longer the responsibility of curriculum experts, but of professional scientists, historians and psychologists supported by massive monetary investments. Bladès (2008) reports that the science curricula, in the 1960s, 1970s and beyond, focused on the professional practice of science and not on an understanding of science. These curricula, Bladès (2008) asserts, were very superficial. He writes:

> Science, in this case, was defined in a superficial, circular way as the professional practice of scientists who methodically discovered through a systematic method 'facts' about how the world operates and then used these facts for the betterment of humankind. The rational-objectivist, positivist assumptions of this approach to science and science education ignored debates about the nature of science. (Bladès, 2008, p. 388)

The motivation behind the design of these curricula was political and economic in order to create more scientists as a way to achieve scientific and technological superiority over the communist nations.

Joseph Schwab (1969) was very critical of these 'theoretical' curricula and advocated for a 'thinking' curriculum in which he discouraged the theory/practice divide and pointed out that curriculum development must start with the personal lived experiences of learners in order to engage them critically. His classical work on 'science as doing' argues that if anyone wants to understand science the focus should be not on what a scientific theorist is saying, but rather on what he or she is doing. Duschl (1988) concurs with Schwab and points out that school science curricula not only should address what is known about science, but should include how scientists arrived at such knowledge claims. To these scholars a healthy and vigorous understanding of science can only emerge if a teacher values and embraces the intensity and fullness of a learner's experience. This is because they believed that experience is the gateway through which scientific meaning can be grasped. According to Schwab (1969), experience does not only give one specific meaning to an event, but allows multiple ways of knowing, seeing, smelling, touching and hearing. In other words the senses become the main scientific instruments with which learners explore, learn and understand their environment.

In Africa and South Africa scholars like Aikenhead (1996), Jegede and Aikenhead (1999) and Ogunniyi (1988, 2007), amongst others, like Schwab in America, have decried the state of school science for decades. Through various empirical studies these researchers stressed the importance of also valuing indigenous knowledge that an African learner brings into the classroom. They argue that if curricula and the teaching thereof do not take cognisance of a learner's social and cultural knowledge, the learning of science is reduced to mechanistic and instrumental activities. Jegede (1999) avers that the goal of a school science curriculum is to guide learners towards effective ways of understanding the processes of science by drawing from their collective engagements with the physical world around them. Failure to do so will lead to 'naïve conceptions', 'misconceptions', 'pseudo science', 'alternative conceptions' and 'children's science' that is often obscure and misaligned with the real theories of science. In the process the learning of science becomes dogmatic, inappropriate and meaningless to learners. This view of science leads to a mono-dimensional way of thinking that constrains rather than emancipates the learners' outlook on science and the world.

According to Le Grange (2016), this mono-dimensionality of curricula still dominates our school science programmes to this day. He reports that all our curricula constrain rather than emancipate our learners. Jegede (1995) substantiates this claim by pointing out that in developing countries the school's science curriculum is a major constraint in developing multidimensionality because learners' strong belief in their communities' indigenous belief system clashes with the way science is presented to them. What is needed to emancipate our learners, Le Grange (2016) echoes, is a curriculum that inspires teachers to tap the rich imaginative powers and creative abilities of their learners. Because such a curriculum will encourage teachers to unearth the rich cultural worldviews of learners as a source to introduce the new content to be learnt. To do so, teachers must plan their classroom activities and pedagogical engagements to develop such capacities. This raises the question: Does CAPS promote a humanistic approach to science to break away from a politically motivated objectives-driven curriculum?

Brief Overview of Curriculum Post-1994

The introduction of C2005, also referred to as outcomes-based education, was a response to the initial period of non-intervention (1994–1997) under the new government of national unity. Prior to its inception under apartheid school science curriculum was often criticised as one that was content heavy and examination driven, and promoted the rote learning of canonical science that was far removed from the learners' personal lived experiences (Jansen, 1999). Consequently, C2005 adopted an outcomes-based approach, underpinned by a constructivist philosophy in which 'outcomes' (skills, knowledge and values) replaced the term 'content'. Spady (1994) defines 'outcomes' as high quality, culminating in demonstrations of significant learning in context. Under this curriculum teachers had to shift their practices to a more learner-centred curriculum in which humanistic science had to be promoted. To do so teachers had to clearly define the learning outcomes for every classroom activity and the lessons had to be designed to reach the desired outcomes.

The shift to C2005 also demanded a change in the pedagogical approaches of teachers which had remained unchanged under the apartheid curriculum for decades, despite new advances made in science over the years. This meant that teachers had to come up with alternative teaching approaches and assessment strategies that encouraged meaningful learning. To Ausubel (1968) 'meaningful learning' takes place when a teacher links what a learner already knows with what he/she ought to or is expected to know. It is the juxtaposition of the learner's everyday realities and experiences with school science that makes science meaningful. As a result of inadequate training, the absence of a formal method on how to implement the new curriculum and a lack of support, teachers struggled with the implementation of C2005. This led to a Curriculum Review process in which substantial changes were recommended by a committee appointed by the Minister of Education Kader Asmal during that period. Consequently, C2005 was replaced with an interim RNCS, which was implemented in 2002. The newly legislated RNCS maintained its core contrasting philosophies of transformation and a neoliberal agenda of high knowledge and skills which formed the crux of the curriculum. However, this curriculum (RNCS) had a stronger focus on content (canonical science) and the broad vision of the curriculum did not connect with the needs of the learners and communities at grassroots level (Harley & Wedekind, 2004). This shortcoming laid the foundation for the Further Education and Training National Curriculum Statement, which was implemented in 2008.

The year 2008 marked the introduction of the first National Senior Certificate examination for all schools in South Africa. Because of the hasty phasing in of the curriculum many teachers complained about being inadequately prepared for effective implementation (Matoti, 2010). Their main concerns were a lack of training, inadequate content knowledge, complex jargon used in the curriculum and a lack of scarce resources such as laboratory facilities, libraries and internet connectivity (for full details see Koopman, 2017). In order to assist teachers further, two additional supporting documents were drafted at national level, that is, the learning programme and a subject assessment guideline (SAG) (DoE, 2008a, 2008b). The SAG helped teachers to prepare learners adequately for the final examination. For example, in Physical Science the SAG provided teachers with a clear focus on what the weightings of each topic in the syllabus would be for the two different disciplines of Physics and Chemistry. This document also gave teachers an idea on the taxonomical classification (Bloom's taxonomy) of the content and examination, and the respective percentages for the learning outcomes (for full details see DoE, 2008b; Koopman, 2010). Furthermore, it also provided teachers with example question papers of what will be tested and how the questions will be structured. Critics continued to level their complaints about this curriculum as being content heavy, confusing and objectives driven, while teachers complained about the complexity of the technical aspects such as learning outcomes, critical outcomes, developmental outcomes and their associated assessment standards. This ultimately led to the appointment of a task team appointed by the new Minster of Education Angie Motshekga in 2009. The final outcome of this review process was that the NCS had to be streamlined by simplifying the curriculum. This led to the introduction in 2012 of the Curriculum and Assessment Policy Statement (CAPS).

Methodology

The study adopted a qualitative approach in which the policy text of the CAPS was analysed. The focus of the analysis was on understanding the various statements and phrases in the CAPS that speak directly to these research questions. Honan (2004) points out that 'teachers engage rhizomatically with policy texts; some adopt, and some resist, some subvert, and so on' (p. 4). This accentuates a multidimensional nature of the reading and interpretation of a policy text, which leads to an understanding of policy that can disrupt or fail to disrupt a teacher's disposition towards his or her teaching method. According to Spivak (1996), there is no one correct path to understanding any text, but it is the relationship between the teacher and his or her connection to the text that can lead to them adjusting their practice. Spivak also points out that the way that the teacher connects with policy text also depends on where the teacher is located within a particular school setting. The aim of this analysis is therefore more concerned with whether CAPS advocates for canonical or humanistic science as well as the teaching thereof. It is important to note that although the policy text is legitimated by law, it remains rules and not laws that guide the teaching profession. This means there is a sharp contrast between rules and laws. This contrast lies in the fact that human behaviour can be predicted based on their knowledge of the law, but we cannot predict behaviour based on rules, because rules are context bound.

Coding of Section 1 of CAPS

Section 1 of the document focused on the general background of the curriculum, such as its aims, objectives, purpose and principles. This section provided an insight into the philosophical framing that drives the Physical Science curriculum. In this section I coded each concept that pointed to the promotion of *canonical science* and/or *humanistic science*. While canonical science is knowledge *about* science such as facts, formulas, calculations, theories, principles and laws a humanistic science is about linking the content to human subjectivity, personal lived world experiences and local knowledge of the community. Each statement, phrase or concept was transferred to an Excel spreadsheet with a researcher's note to elucidate its action as stipulated in the policy. To do this I searched for key words such as 'knowledge', 'skills' and 'values' within the two frames of reference to determine the emphasis placed on each of these focus areas, as this has implications for the teacher and his or her role in the classroom. I attempted to determine the context in which phrases and concepts such as knowledge, skills and values occurred. I also made a list to separate the canonical focus on science reflected in the text and compared it with a more humanistic approach to science. I excluded all the general information, tables and glossaries, as these are repetitions of textual information.

Analysis of Section 2 of the CAPS

Section 2 of the analysis focused mainly on how the subject is described or defined, its 'specific aims' (which replaced the outcomes espoused in the NCS), the knowledge areas (which replaced the knowledge strands in the NCS) and the taxonomy of the various assessment tasks. I summarised most of the information in this section in table format, as some information is repeated. Here the important aspects considered were the 'specific aims', which form the crux of the pedagogical approaches, and the knowledge strands, which form the core of the curriculum. These aspects provide an insight into the curriculum framework that defines the approaches taken by teachers when they engage with the curriculum in the classroom inter-phase. For example, in the description or definition of Physical Science the focus is on scientific investigation and inquiry and on indigenous knowledge, but very little attention is devoted to indigenous knowledge in the core content.

Analysis of Sections 3 and 4 of CAPS

Section 3 was analysed in detail as it focused mainly on the content, the time allocated to cover each topic, key concepts, suggestions for the practical activities, resources required and teacher guidelines to direct the teachers' focus for each topic. This section constituted 127 pages of the entire 164 page document. This section was coded for details on the emphasis placed on *canonical science*, which promotes rigid, mechanistic and socially sterile content, and *humanistic science*, which resonates with personal lived experiences of learners. To do this two columns were constructed on an Excel spreadsheet to differentiate between canonical and humanistic science, and entries were made under each column by searching for key words. For example, under canonical science the focus of the analysis was on key words such as *give, name, identify, define, relate, calculate, write* and so forth. For humanistic science the focus was on key words and phrases such as 'apply within context', 'awareness', 'indigenous knowledge', 'folklore', 'demonstrations within context' and the use of everyday products for scientific scrutiny and so forth. Any statement, phrase or concept in CAPS that pointed to either of these two focus areas was entered on the spreadsheet. After completing the analysis of the content in Section 3, I counted how many entries were made under each column for all grades, after which I separated the entries to ascertain the relationship between canonical and humanistic science for every grade in the phase.

Section 4 outlined the types of assessment, the weightings and the taxonomical classifications. All these sections were coded and summarised to determine what types of assessments were required by the teacher, number of tasks per grade to determine the number of way and appropriate way. This also provided an insight into the pedagogical approaches that a teacher might follow to ensure that he or she complies with the requirements stipulated in CAPS in order to reach the desired objectives. I present a synopsis of the findings below.

Results

The analysis of the CAPS was undertaken to provide answers to the following two main research questions: (i) Does CAPS for Physical Science promote 'canonical' or 'humanistic' science and its associated knowledge? (ii) How do this approach to science and its associated knowledge promoted in CAPS (re)configure the pedagogical practices of teachers? The findings consequently focus only on these two questions.

Section 1 of CAPS: General Aims and Objectives Pertaining to Physical Science

The general aims of CAPS are divided into three sections: a general statement, the purpose of the curriculum and the principles according to which the curriculum is designed. In the opening statement the emphasis is on the 'knowledge, skills and values' that are meaningful and applicable and contextualised within the learners' experiences. This is substantiated by the fourfold purpose statement of CAPS, the first of which emphasises the importance of the knowledge, skills and values necessary for 'self-fulfilment' so that 'meaningful' participation can take place in the respective societies. The remaining three purpose statements devote attention to 'access to higher education', 'education for the workplace' and 'providing employers' with competent learners'. These three purpose statements clash with the first statement on the 'knowledge, skills and values' for self-fulfilment, as the latter refers to humanistic science, the former speaks to canonical science that is factual, mechanistic and technical in nature. A curriculum that promotes canonical science could stifle the learners' humanistic development as its objectives revolve around the preparation of learners for the labour market which is synonymous with the objective of the apartheid curriculum that aimed at constraining instead of liberating a learner's outlook and understanding of the world.

CAPS is based on seven principles: 'social transformation', 'active and critical learning', 'high knowledge and high skills', 'progression' (content and context of each grade from simplex to complex), 'human rights, inclusivity, environmental and social justice', 'Valuing indigenous knowledge' and 'credibility, quality and efficiency' (comparing the quality of education in South Africa with that of other countries). While social transformation, human rights and indigenous knowledge speak to the promotion of humanistic materials and approaches to science teaching, according to which the learner and his or her lived world should be placed at the centre of the educational agenda, the remaining four principles are mainly concerned with a neoliberal agenda. This agenda drives canonical science with a focus on science as an academic discipline that aims to prepare learners for the labour market. In essence it underlies a teacher-centred approach to curriculum with more emphasis placed on concepts, principles, laws and theories of science delivered in a rigid and instrumental way.

The type of learner the curriculum envisages producing is a critical, creative thinker who can work effectively in a team in order to become responsible citizens. To achieve this, teachers must assist learners in the collection, analysis and organisation of data so that they can view science as a practice of inquiry that critically evaluates phenomena. When this happens, learners also develop the skill to use science and technology in a responsible way. The term 'critical' is used three times to refer to the type of learner, which raises the question of whether a strong canonical focus on the teaching of science can nurture critical thinking in a classroom environment characterised more by an economic agenda as opposed to a humanistic agenda. Out of 21 statements coded in this section 17 (80.9%) focus on canonical science which requires dogmatic science instruction, while 4 (19.04%) emphasise humanistic science which is meaningful and appropriate to the learners' lived world, enabling them to develop a multidimensional outlook on science and the world. Johnson, Dempster, and Hugo (2015) in their analysis of the CAPS for Life Science report similar findings and found that the curriculum favoured canonical science more and devoted very little attention to humanistic science. In their study they coded 422 statements throughout the document, of which 296 (70%) were coded as canonical materials and 126 (29.9%) were coded as humanistic materials.

Section 2: The Specifics of Physical Science

On the one hand CAPS defines Physical Science emphasising the importance of both 'scientific investigation' and 'inquiry' (canonical science), while on the other hand it stresses the importance of 'indigenous knowledge'. Table 1.1 lists the three specific aims of teaching Physical Science:

Table 1.1: Specific aims of CAPS for Physical Science

Specific aim: 1	Scientific inquiry and problem-solving skills
Specific aim: 2	Construction and application of scientific knowledge
Specific aim: 3	Preparing learners for careers in science, to be responsible citizens and to appreciate the environment

Source: Author.

The way in which the 'specific aims' (SAs) are sequenced suggests a practical approach in which learners should discover and construct their own understanding and knowledge of science through practical scientific investigations. This is captured in SA 1, which forms the basis for every scientific problem which should be addressed through *scientific inquiry*. SA 1 feeds into SA 2 in which the *doing aspect* of science (SA 1) leads to the construction of scientific knowledge (SA 2) or '*knowing science*' so that learners can develop a conception of its applications to real-life situations or contexts (SA 3).

Section 3: Analysis of the Prescribed Content

In this section a total of 670 entries were made across all three grades (10–12). Out of the 660 entries 603 (91.36%) represented theoretical or canonical science, and only 57 (8.63%) were humanistic science that is related to the lived experiences and 'local' knowledge of the learners. Table 1.2 gives a breakdown per grade of all the entries with respect to canonical science versus humanistic science that focuses on real-life events and personal lived experiences of the learners.

Table 1.2: Comparative weightings of canonical and humanistic science in the FET syllabi

Grades	10	11	12	Total	Percentage (%)
Canonical science (theory)	218	202	183	603	90.55
Humanistic science (real-life events/experiences)	19	36	2	57	9.45
Total	237	238	185	660	100

Source: Author.

The results show that in Grade 10 a total of 237 entries were made, of which 91.98% of the content placed more emphasis on canonical science, while 8.02% were humanistic science by focusing on the real-life experiences of learners. In Grade 11 a total of 238 entries were made, of which 84.87% placed more emphasis on canonical science with its theoretical knowledge, while 15.13% of the content devoted attention to humanistic science. In Grade 12 only 1.09% of the content hinged on humanistic science, whereas 98.91% focused on canonical science. The results show that Grade 12 focused more on canonical science compared to Grades 10 and 11. This is understandable given the high priority placed on the Grade 12 results nationally. Figure 1.1 presents the findings in the form of a bar graph.

Figure 1.1: Findings in the form of a bar graph

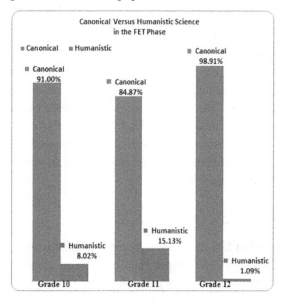

Source: Author.

In Grades 10 and 11 a total of 120 hours are devoted to teaching time and practical work, while 40 hours per grade per year is dedicated to examinations and tests. In Grade 12, 104 hours are dedicated to contact time for teaching and practical work, while 14 hours are dedicated to examinations. The results show that the CAPS syllabus (content) across all the grades focuses more on canonical science that does not empower teachers to use innovative teaching strategies that could emancipate their learners from the constraints of mechanistic, instrumental and rote learning of canonical science. Given this strong focus on canonical science across all the grades in the FET phase, it seems far-fetched for teachers to introduce their learners to a different view of science in which lived experiences become the focus of their class-room practices. This means teachers could be limited in their attempts to transform their practices, because of a predetermined set of ideas derived from a specific body of knowledge, in order to liberate their learners. This prescriptive body of knowledge across Grades 10–12 makes visible the power lines and networks that control and shape the pedagogical practices of teachers. Furthermore, it stresses the rigidity of an objectives-driven Tylerian model of education that is counter-intuitive to learning science through the personal lived experiences of learners placed on the periphery.

Section 4: Assessment for Physical Science According to CAPS

This section focuses on the programme of assessment for each grade; this involves four steps: generating and collect evidence, evaluation of the evidence, recording the findings and using the findings to assist learners in the learning process. CAPS distinguishes between informal and formal assessment. Informal assessment is de-scribed as the teacher's daily engagement with his or her learners in the classroom, which involves observations, discussions, practical demonstrations and so forth. These types of assessments do not contribute towards the learners' programme of assessment for grading or certification. Formal assessment, on the other hand, contributes towards administration for promotion and certification processes. It includes tests, examinations, practical investigations and research projects or as-signments. Table 1.3 summarises the process.

Table 1.3: Overview of formal assessments in the FET band

Assessment	**Grade 10/11**	**Grade 12**
Practical work Assignment/Research projects Tests	Four prescribed practicals per year One practical/term: Physics or Chemistry and One project/year in Physics or Chemistry Two control tests per year One mid-year exam (Weighting = 25%)	Three prescribed practicals per year; One or two Chemistry or one or two Physics practicals No project One control test per year One mid-year exam One trial examination (Weighting = 25%)
Final Assessment	Two examinations per year (Weighting = 75%)	One final external examination (Weighting = 75%)

Source: Author.

These formal assessment tasks are specified to be age and development specific, and related to the topics covered in syllabus. This means the teacher must set a variety of different tasks for each topic to meet the needs of every learner. All practical investigations, including experiments, research projects or assignments, carry a weighting of 25% towards the final mark, while control tests/mid-term examinations and final examination contribute 75% towards the final mark. All the tests and examinations (Grades 10–12) must adhere to the taxonomical classifications listed in Table 1.4.

Table 1.4: Taxonomical classification of tests and examinations

Cognitive level	Description	Paper 1 (Physics)	Paper 2 (Chemistry)
1	Recall	15	15
2	Comprehension	35	40
3	Analysis and application	40	35
4	Evaluation and synthesis	10	10

Adapted from: DBE (2011).

Most of the cognitive level of difficulty lies between levels 2 and 3, which make up 75% of the taxonomy for both chemistry and physics. While level 2 tests for understanding require the learner to interpret and relate the information that he or she learned to the content, level 3 is more concerned with a higher cognitive order in which the learner not only integrates the content into a particular context, but is also able to break down a particular scenario into its scientific concepts, theories and laws. This involves key words such as deconstructing, comparing, structuring and so forth. Level 1 constitutes 15% of taxonomical classification, while level 4 constitutes 10%. While level 1 focuses on naming, identifying or remembering facts, level 4 requires the learner to work with new ideas that are abstract in nature, which he or she must reformulate or compile evidence to substantiate a particular hypothesis.

At the end of each year the teacher must compile a portfolio of evidence containing all the assessment tasks, assessment instruments, rubrics and memorandums for marking, evidence of learners' work as well of a summary of the marks on which the learners have been graded. The portfolio of evidence for Grades 10 and 11 is internally moderated, while Grade 12 portfolios are externally moderated.

Discussion

What emerges from these findings is a national curriculum framework that is content-focused and assessment-driven, given all the emphasis placed on canonical science and assessment. Pedagogical work that aims to emancipate learners requires an engagement with learners that starts from a specific point of development and is cognisant of the horizon of understanding of learners (Shulman, 2005). But given the overwhelming focus on canonical science and the marginalisation of humanistic science with its lived world approach this is not the case. The 'huge emphasis placed on continuous assessment (CASS) alienates teachers and distracts them from more interactive pedagogies' (Beets, 2012, p. 70). No substantive unpacking of a productive pedagogy is provided for how teachers can deliver the content to a culturally diverse learner community and no link is made between pedagogy and assessment. Rather than elaborating on how to design quality assessment activities, so that teachers can produce valid evidence on quality teaching and learning, the document focuses more on content, the various types of assessment and the number of assessment tasks required in every grade.

According to Le Grange (2016), these 'specific aims' have more in common with Tyler's (1949) approach to the curriculum, given the fact that it has become more prescriptive regarding the 'what', 'when' and 'how' teachers must teach. Tyler's approach to the curriculum is an objective-driven, tightly packed factory model in which schooling is organised so that children must be guided through schooling by identifying specific goals that they must achieve. These goals are driven by regular assessment to determine whether the learner has achieved the objectives set by the specific aims. This curriculum model is based on industrial rationality, where the learner is viewed as a consumer of knowledge. Davis (2006) notes that teachers become preoccupied with continuous testing for good results and thereby do not apply effective pedagogical strategies to connect the rich content to its applications, so that learners can use their knowledge intelligibly and flexibly in diverse circumstances in real life. Koopman (2013) and Koopman, Le Grange, and de Mink (2016) report similar findings that teachers teach to test in order to obtain good results in Grade 12, because of the pressure placed on teachers by government.

As an aspect of the control over the teachers' work, teachers (Grades 10–12) have to comply with the policy requirement to provide teacher portfolios of assessment tasks and assessment instruments, evidence of learners' work as well as a summary of marks used to grade the learners. Beets (2012) points out that this interferes with the day-to-day activities of how teachers do their work, because this means that their own performativity in terms of the state's performance indicators and their accountability to their respective authorities are given a higher priority than the interests of their learners, the parents and the society. Blasé and Anderson (1995) refer to this as the dynamics of subordination which, they argue, results in unauthentic teaching behaviour based on the need to survive. Consequently, teachers contribute to their own subjugation and de-professionalisation by complying with policies of control rather than seeking for new ways or strategies to exert their agency in the learners' interests. All of these restrictions and responsibilities placed on the teachers' doorstep means that the pendulum under the current CAPS has swung to the side of teacher accountability and performativity with little regard for learners and their pedagogical needs. The situation is further exacerbated by principals who act as line managers in a factory, further pressurising teachers for good results as a result of the pressure placed on them by their superiors (Ball, 1993).

When a curriculum places too much emphasis on canonical science and assessment, a teacher's role becomes predetermined while hampering the intellectual well-being of the learner. In other words, the role of the teacher is to deliver predesigned accounts of knowledge that is meaningless in, and inappropriate for, their lived world. Arendt (1993), a humanist of note, holds that in a capitalist society the education of children is in danger because of the imperatives of a political and economic system that reduces action to behaviour through mechanisms such as the homogenisation of learning under a regime of high-stakes tests and examinations. This testing and examining Arendt refers to as absolute control in which behaviour and conduct become predetermined, controlled and measured. She asserts that teachers should resist, intervene and contradict policy in order to restore the freedom of the child as an active thinking being, in order to make them aware of their surroundings. She contends that the child should be reared in a way that cultivates and preserves them for future generations.

Conclusion

The findings in this study have shown that CAPS is an objective-driven, content-laden and assessment-dominated curriculum that promotes canonical science. The findings have also shown that a focus placed on assessment discourages rather than encourages humanistic science, in which the learner's personal lived world experiences form the centre of the teaching and learning process. This means that instead of the teacher producing liberated learners through pedagogies of engagement that foster critical thinking, teaching becomes reduced to a technical-rational and mechanistic focus on canonical science. Furthermore, the prescriptive nature of the curriculum also forces teachers to become textbook bound, which in turn promotes knowledge that is inappropriate and meaningless to the lived world of the learners. Additionally, the increased pressure on global capitalism increases the focus of the curricula on the 'ends' rather than the 'means'. In the process the performance of the learners is measured by how closely they follow the prescribed rules, while the sequences of how the content must paced and methodised are laid down externally.

According to Ayers (2006), the core of a teacher's work should be to generate human knowledge and human freedom, which is both enlightening and emancipatory for learners. To do this requires a curriculum that empowers a teacher to act accordingly and does not suppress their voices and efforts. Ball (1993) argues that the curriculum is guided by policy, and policy is a form of surveillance through which governments exercise control and power over the teacher and the learner. This is true, he argues, because the curriculum determines the teacher's work, which is often driven by technical factors that reduce the teacher's professional autonomy. Furthermore, the teacher and the learner are both absent figures in the discourse of policy formulation. As Ball (1993, p. 108) rightly argues, it is not only a question of *what* is said but also of *who* is saying it when it comes to the policy text, as it is an important indicator of what is at stake (p. 109). This 'micro-technology of control', as Ball terms it, is articulated by patterns of self-regulation (p. 111). When teachers express their resistance to policy issues, it works against them rather than against the policy and is often seen as an attack on the survival of the institution. Individuals in management positions, who are often viewed as having more power than the teacher and the learner, are also under surveillance, but they soon become docile in the process and take ownership of the policies. These principals (managers) and their superiors are managed by policies, which Foucault (1991) describes as constituting a 'microphysics of oppression' (p. 112).

References

Aikenhead, G. S. (1996). Science education: Border crossing into the subculture of science. *Studies in Science Education, 27*, 1–52.

Arendt, H. (1993). *What is freedom in between past and future*. New York, NY: Penguin.

Ausubel, D. (1968). *Educational psychology: A cognitive view*. New York, NY: Holt, Rinehart & Winston.

Ayers, W. (2006). Trudge toward freedom: Educational research in the public interest. In G. Ladson-Billings & W. F. Tate (Eds.), *Education research in the public interest: Social justice, action and policy* (pp. 81–97). New York, NY: Teachers College Press.

Ball, S. J. (1993). Education policy, power relations and teachers' work. *British Journal of Educational Studies, 30*(12), 106–121.

Beets, P. (2012). Strengthening morality and ethics in educational assessment through Ubuntu in South Africa. *Educational Philosophy and Theory, 44*(2), 70–87.

Berman, P., & McLaughlin, M. W. (1976). Implementation of educational innovations. *The Educational Forum, 40*, 345–370.

Bladès, D. (2008). Positive growth: Developments in the philosophy of science education. *Curriculum Inquiry, 38*(4), 387–400.

Blasé, J., & Anderson, G. L. (1995). *The micropolitics of educational leadership: From control to empowerment*. London: Cassell.

Bobbitt, J. F. (1918). *The curriculum*. Boston: Houghton Mifflin.

Chinn, P. W. (2007). Decolonising methodologies and indigenous knowledge: The role of culture, place and personal experience in professional development. *Journal of Research and Science Teaching, 44*(9), 1247–1268.

Davis, A. (2006). The measurement of learning. In R. Curren (Ed.), *A companion to the philosophy of education*. Oxford: Blackwell.

Department of Basic Education. (2011). *Curriculum and assessment policy statement for further education and training Physical Science*. Pretoria: Department of Education.

Department of Education. (2008a). *Subject assessment guidelines for FET (Grades 10–12) Physical Science*. Pretoria: Author.

Department of Education. (2008b). *Program of assessment for FET (Grades 10–12) Physical Science*. Pretoria: Author.

Dewey, J. (1970). *Creative intelligence, "A recovery of philosophy."* New York, NY: Octagon Books.

Doll, W. E. (2008). Complexity and the culture of curriculum. *Educational Philosophy and Theory, 40*(1), 190–214.

Duschl, R. (1988). Abandoning the scientistic legacy of science education. *Science Education, 72*(1), 51–62.

Foucault, M. (1991). Governmentality. In P. Rabinow (Ed.), *The Foucault reader: An introduction to Foucault's thought* (pp. 1–22). Harmondsworth: Penguin.

Hamilton, D. (2003). *Instruction in the making: Peter Ramus and the beginnings of modern schooling*. Paper presented at the 2003 annual meeting of the American Educational Research Association.

Harley, K., & Wedekind, V. (2004). Political change, curriculum change and social transformation, 1990 to 2002. In L. Chisholm (Ed.), *Changing class: Education and social change in post-apartheid South Africa* (pp. 95–220). London: Zed Books.

Honan, E. (2004). (Im)plausibilities: A rhizo-textual analysis of policy text and teachers' work. *Educational Philosophy and Theory, 36*(3), 267–283.

Jansen, J. (1999). Setting the scene: Historiographies of curriculum policy in South Africa. In J. Jansen & P. Christie (Eds.), *Changing curriculum: Studies on outcomes based education in South Africa*. Kenwyn: Juta.

Jegede, O. (1995). Collateral learning and the eco-cultural paradigm in science and mathematics education in Africa. *Studies in Science Education, 25*, 97–137.

Jegede, O. (1999). Science education in nonwestern cultures: Towards a theory of collateral learning. In L. Semali & J Kincheloe (Eds.), *What is indigenous knowledge? Voices from the academy (119–142)*. New York: Falmer Press.

Jegede, O., & Aikenhead, G. (1999). Transcending cultural borders: Implications for science teaching. *Journal of Science Teaching and Technology, 17*(1), 45–66.

Johnson, K., Dempster, E., & Hugo, W. (2015). Exploring the recontextualisation of biology in the CAPS for life sciences. *Journal of Education, 60,* 101–122.

Koopman, O. (2010). *An analysis of the first FET Physical Sciences examination.* Paper presented at the annual Southern African Association for Research in Mathematics, Science and Technology Education. Hosted by the University of Kwazulu-Natal, Durban.

Koopman, O. (2013). *Teachers' experiences of implementing the further education and training science curriculum.* Unpublished doctoral dissertation. Stellenbosch: Stellenbosch University.

Koopman, O. (2017). *Science education and curriculum in South Africa.* New York, NY: Palgrave Macmillan.

Koopman, O., Le Grange, L., & de Mink, K. (2016). A narration of a Physical Science teachers' experiences of implementing a new curriculum. *Education as Change, 20*(1), 149–171.

Le Grange, L. (2016). *Rethinking learner centred-education: Challenges faced by the African child when learning school science and maths.* Unpublished paper.

Matoti, S. N. (2010). The unheard voices of educators: Perceptions of educators about the state of education in South Africa. *South African Journal of Higher Education, 24*(4), 568–584.

Ogunniyi, M. (1988). Adapting western science to traditional African culture. *International Journal of Science Education, 10*(1), 1–9.

Ogunniyi, M. (2007). Teachers' stances and practical argument regarding a science – Indigenous knowledge systems: Part 1. *International Journal of Science Education, 29*(8), 963–986.

Pinar, W. (2004). *What is curriculum theory.* London: Lawrence Erlbaum Associates Publishers.

Pinar, W. (2015). *Educational experience as lived.* New York, NY: Routledge.

Schwab, J. J. (1969). The practical: A language for curriculum. *School Review, 78,* 1–23.

Shulman, L. (2005). *The signature pedagogies of the professions of law, medicine, engineering, and the clergy: Potential lessons for the education teachers.* Paper delivered at the Math Science Partnership Workshop. Hosted by the National Research Council's Centre for Education. Irvine, California.

Spady, W. G. (1994). *Outcomes based-education: Critical issues and answers.* Arlington: American Association of School Administrators.

Spivak, G. (1996). Explanation and culture Marginalia. In D. Landry & G. MacLean (Eds.), *The Spivak Reader. Selected works of Gayatri Chakravorty Spivak.* New York, NY: Routledge.

Taylor, F. W. (1911). *The principles of scientific management.* New York, NY: Harper & Brothers.

Tyler, R. W. (1949). *Basic principles of curriculum and instruction.* Chicago: University of Chicago Press.

CHAPTER TWO

How to Teach Western Science and Indigenous Knowledge as Complementary Bodies of Knowledge

Introduction

Traditionally, in South Africa the doctrine of Fundamental Pedagogics (FP), juxtaposed with the ideology of Christian National Education (CNE), inclined teachers to blame their learners if they did not understand the subject matter and failed their tests and examinations (FP and CNE will be explained in more detail in Chapter 3. For a full account see Koopman, 2013; Krüger, 2008; Le Grange, 2008). Drawing from my personal experiences as a learner during that time, I can clearly remember how many of my classmates often had to suffer the humiliation of being labelled as 'lazy', 'stupid' and 'dumb', simply because they could not answer a fundamental question. Instead of our teachers recognising and questioning the social forces, innate (in)abilities and cultural differences of their learners, one can clearly see that they contributed enormously to the problem of poor science teaching and became the very reason why many learners did not like science or chose to pursue a career in science. Koopman (2017) writes:

> ...our teachers create and instil inexplicable states of consciousness characterised by fear, vilification, illogical trains of thought and other phobias into our minds...*creating* opposition to the prevailing traditions in class and at school—such learners were flogged with a cane in class or sent to the principal's office for more severe forms of punishment. (p. 58)

Back then (and today still) many teachers 'blamed the victim' and restated the problem in terms of the erroneous 'knowledge', 'values' and 'beliefs' the learners brought into the classroom. This reminds me of how my science teacher would constantly say to us in a very aggressive and scary tone: 'In science we don't believe in the nonsense you learn at home and bring to school. In this class we are only interested in hard core facts.' Today, through the proliferation of published works, educational policies and curriculum documents teachers are encouraged to think differently about their teaching and relationship to their students by examining more carefully their own assumptions, beliefs, practices and their actual effects on learners (Koopman, Le Grange, & de Mink, 2016; Ogunniyi, 1987, 2007). Despite continued growth in learner diversity, research has shown that many teachers are still trapped in traditional pedagogies in which learners are encouraged to slavishly follow and confirm teacher decision making in practical work and mainly focus on theory in the science classroom (Hodson, 1993; Mji & Makgato, 2006). Seltzer-Kelly (2013) adds that, underlying this approach and magnifying its effects is the failure by teachers to understand the fundamental nature of pedagogy—which, in a narrow sense, means how to teach.

Often teachers cannot connect with their learners because they are socially, culturally and epistemologically disengaged from the lived world experiences of their learners. Consequently, teachers find themselves epistemologically located in some fixed subjectivity that positions them in such a way that all they see before them in the classroom are empty vessels to be filled. In other words, they don't see their learners in their fullness as unique, living, subjective and epistemologically active beings who are constantly engaged with the physical world. Hall (1989, p. 190) refers teachers to the importance of being engaged with a child's past and history in the teaching and learning space when he writes:

> The past is not only a position from which to speak, but it is also an absolutely neces-sary resource in what one has to say.... Our relationship to the past is quite a complex one; we can't pluck it up out of where it was and simply restore it to ourselves.

In agreement with Hall's point, many African scholars have over the years shown the powerful impact of the historical, social and cultural forces on the way that African learners think and view the world. How these forces culminate in school science curricula encouraged these scholars to scrutinise the various science cur-ricula in Africa and South Africa carefully and critically not only for discrimina-tory core content, but also for the absence of effective pedagogies (for details, see Aikenhead, 1996; Jegede, 1989, 1999; Jegede & Okebukola, 1991; Le Grange, 2007; Naidoo & Lewin, 1998; Ogunniyi, 1987, 1988). Their research findings have shown that the main contributing factor to this dilemma is the canonical Western tradition that continues to dominate the subject.

For many indigenous populations this image (culture) of science is a colonising one, because most African learners cannot connect with this type of science and the way it is being taught. Furthermore, the prominence of Western-oriented science in our school science curriculum, with its canonical approach to teaching, promotes the view that African indigenous knowledge is non-scientific and confined exclusively to the supernatural (Shizha & Emeagwali, 2016). Aikenhead (2008) accurately summarises the research findings on the views of learners about school science when he writes that many learners describe school science as being 'socially sterile, impersonal, frustrating, intellectually boring, and dismissive of their life worlds and career goals' (p. 26). This statement is justifiable given the fact that traditional culture forms the dominant part of many African learners' episteme (Fadkudze, 2004), which means the knowledge they hold about the world is essentially related to their human nature. Discounting and underrating these epistemological and ontological dispositions of learners towards school science conveys the perception that indigenous knowledge is irrelevant to their daily realities, questionable and false. Whether it is mathematics, astronomy, medical science or agriculture, this irreplaceable knowledge faces extinction unless it is infused into the school science curriculum and plausible teaching approaches and strategies are designed to facilitate its smooth transmission from one generation to the next.

After the advent of South Africa's newly found democracy in 1994, policy-makers and curriculum planners decided to introduce indigenous knowledge into the school science curriculum. The rationale for making it part of the school science curriculum was outlined as follows:

> The prevailing world-view of science is based on empiricism…However, the existence of different world-views is important for the Natural Sciences curriculum. One can assume that learners in the Natural Sciences Learning Area think in terms of more than one world-view. Several times a week they cross from the culture of home, over the border into the culture of science, and back again…These South African issues create interesting challenges for curriculum policy, design, materials and assessment. (Department of Education [DoE], 2002, p. 10)

Although curriculum policy-makers and planners intended to pass on this rich repository of local knowledge to all African learners, Ogunniyi (2007) reports that under C2005 the incorporation of indigenous knowledge into the school science curriculum failed abysmally because (i) teachers were trained in the tradition of Western science and were not familiar with indigenous science; and (ii) teachers had resisted change to their pedagogical strategies in terms of contextualisation and indigenisation. Rather than focusing on effective teaching strategies that gave due recognition to both Western and indigenous knowledge, teachers placed greater emphasis on learners mastering scientific information for examinations. Despite these challenges facing the successful implementation of indigenous knowledge under the NCS, South Africa's latest Curriculum and Assessment Policy Statement (CAPS) for Physical Science (Department of Basic Education [DBE], 2011), which is a modification of the National Curriculum Statement, continues to give recognition to indigenous knowledge as it forms part of the country's definition and description of Physical Science as a school subject. In its two-part definition CAPS defines Physical Science as, on the one hand, a practice of inquiry to investigate both chemical and physical phenomena, but adds, on the other hand:

> The subject also deals with society's need to understand how the physical environment works in order to benefit from it and responsibly care for it. All scientific and technological knowledge, including Indigenous Knowledge Systems (IKS), is used to address challenges facing society. (DBE, 2011, p. 8)

However, one of the major concerns of this curriculum (CAPS) is that, although part of the description of the subject includes indigenous knowledge, very little consideration is given to indigenous content and how to teach it (IKS). This means that indigenous knowledge is not getting the same attention in the core content of CAPS as canonical Western science. This chapter therefore asks the question: What teaching strategies can teachers use to critically engage their learners by giving due recognition to both Western and indigenous knowledge systems in order to resolve the tension between these two competing/complementary bodies of knowledge? Given the ongoing debate on and obsession with questions of 'what knowledge' and 'whose knowledge' are of the most worth to the people of South Africa, this study offers teachers possible ways of teaching science at the level of practice when it intersects with both worldviews. In an attempt to answer this question, I will first provide a succinct differentiation between Western and indigenous knowledge systems. Second, I discuss why it is important to incorporate indigenous knowledge into the school science curriculum. To do so, I draw from Merleau-Ponty's (1962) theory of 'the lived body' to explain that learning is an embodied process in which the 'sum total parts of the body' are connected in the learning of science. Thirdly, I use Aikenhead's (1996) theory of border crossing and Jegede's (1999) collateral learning theory to explain the mindset of African learners and possible challenges that they might face when a form of science is taught that is far removed from their cultural disposition. Lastly, I will propose instructional approaches and examples that science teachers can use in the teaching and learning of science, giving due recognition to both Western and indigenous scientific systems.

Differentiating Between Western and Indigenous Knowledge

The debate in education about 'whose knowledge is of most worth' is part of a much wider contest between indigenous and Western forms of knowledge, but more importantly, rival claims about which one is superior to the other. The relevance of this debate is underpinned by the fact that Western science both informs the core content of school science curricula across the world and how science should be taught. According to Ogunniyi (1988), indigenous knowledge is a system of thought generated from ideas that are anthropomorphic, monistic and metaphysical in nature, and that is 'local' or specific to a particular place. This knowledge is informed by a latent worldview that drives the everyday actions and behaviour of many African people. This worldview or mindset guides cultures in their everyday engagement with the world (Le Grange, 2004).

In the Western tradition knowledge is viewed as canonical and mechanistic in nature, as it 'uncouples' or separates knowledge from place or its location and suggests a more 'universalist' view of the world (McKinley, 2005). In order for knowledge to be classified as valid Western knowledge it must be empirically verifiable, must conform to ontological principles, theories and laws, and must be experientially consistent. It is argued in the literature that this universalist perspective of knowledge with its 'universal essence' is epistemologically regarded as superior and more powerful than indigenous knowledge with its pluralist view of the world (Loving, 1995). This view of knowledge diminishes indigenous knowledge as inadequate and inferior, and therefore it is argued that it should not form part of school science curricula.

Each of these worldviews operates within a unique knowledge framework. Aikenhead (2008) distinguished between these knowledge frameworks by using two Greek terms, namely *episteme* and *phronesis*. Western science is associated with the term *episteme* or theoretical knowledge, which is often disconnected from the knower. Indigenous knowledge is more associated with *phronesis*, which represents ways of living or pluralist knowledge generated from man's engagement with the world. According to Aikenhead (1996), each of these knowledge types is governed by borders that are very porous as people often find themselves in between the borders of one of these two, which can be crossed depending on how embedded the person is in a particular worldview. The challenge is for science teachers to deliver the content of school science that will not alienate the learner from the indigenous worldview. How this crossing between borders takes place will be explained later in the chapter, but I first want to present an argument for why indigenous knowledge is important when teaching science.

Why Indigenous Knowledge Is Important in School Science

Merleau-Ponty's Lived Body Theory

Non-Western learners are introduced to science for the first time when they enter the school system (Le Grange, 2007). Le Grange argues that this introduction of science poses a major challenge to learners as they are introduced to a new world-view that is culturally diverse in which they are expected to think differently. This new world in which the learner is submerged to a large extent is corporatised and commercialised based on the idealisation of the global entrepreneurial class that promotes Western science. The main problem starts when science is taught to African learners in a language that uses terminology and concepts that are foreign and complex to their vocabulary. This way of science teaching reduces the world of the learner to a set of incomprehensible constructs that are disconnected from his or her everyday experiences.

According to Merleau-Ponty (1962) learning cannot be treated as the assimilation of propositional content and inferences drawn from it by a person that is wholly detached from events or experiences of others. Instead, he argues, for learning to be meaningful an incoming stimulus must be received by the sense organs of a person that is connected to a particular object or experience in the world. He writes:

> The objective world being given, it is assumed that it passes on to the Sense-organs messages which must be registered, then deciphered in such a way as to produce in us the original text. Hence we have in principle a point-by-point correspondence and constant connection between the stimulus and the elementary perception. (Merleau-Ponty, 1962, p. 7)

At a physiologically level an incoming stimulus causes innumerable afferent neurons to transport the received stimulation to the central nervous system (CNS). An incalculable number of motor neurons then conduct the stimulus from the CNS to the effectors (a muscle or gland), resulting in a reaction. These sensations and reactions are transferred through the various neurons which connect the nervous tissue. Information is then sent through a series of excitation points through motor mechanisms. In other words, the body activates receptor apparatus that in turn activates a number of autonomous circuits. According to Merleau-Ponty (1962), this represents a coordinated action between the *sum total of parts of the body* connected to objects in space. In other words, the body is not viewed as a passive agent that reacts and adjusts itself to external stimuli which act as causes, but is actively engaged (body) in the learning process. This is not the case with the way scientific material is pedagogically delivered to indigenous learners in South Africa. Instead it is presented pedagogically in a Western tradition as a set of facts linked only to pre-determinate internal mental representations which disregard the body as an experiential entity.

When science is presented as 'localised' knowledge, learning takes place through the *sum parts of the body*. This means both external representations of knowledge (all the senses) and internal representations (neurological actions) connect which like a projecting instrument helps the learners to perceive things/objects with the sum part of the body (Merleau-Ponty, 1962). What Merleau-Ponty alludes to is, in a real experience or event, a phenomenon or object is personally captured by the retina, travels from the eye to the occipital lobe (at the back of the brain) where it is processed. From there it travels via synaptic highways to the frontal and temporal lobes of the brain for meaning and significance. At this point it is retained within the framework of all other connected sense organs which provides the learner with additional information from which the learner develops or leads to a conceptual pre-determinate understanding. Thus, a real event brings the learners external stimulus in line with the constancy hypothesis that clarifies and validates scientific statements, facts and laws under investigation. But what do I mean with a real experience and why does it enhance learning?

According to Jaspers (1997) a real experience is not imagined or fictitious (or theory as in science), but a physical experience that is impressed on the mind as felt space (or awareness). This felt space is retained in the mind as a memory, because a real experience is stored as an episodic memory in the brain in which the significance of each part of the episode/event is physically captured by all the sense organs for processing. When this happens the stage on (or space within) which the event or experience has taken place changes from physical space to mental space. This shift from physical to mental space leads to deeper levels of understanding that results in an outcome of felt knowledge. Thus, knowing is not the accumulation of information, but a recollection of events that occurred in past times and spaces conceptualised as embodied knowledge.

In bringing this discussion closer to the value of indigenous school science means that the world of the learner should be the start and endpoint in the teaching and learning environment. When this happens the learners' understanding of science becomes connected to his or her lived world experiences. The learner's awareness of his or her lived experiences generates a tangible, active model (practical) of engagement with the content. Because Western science dismisses the lived world experiences of learners they feel committed to memorise incomprehensible and disconnected constructs. Consequently, learners behave, in metaphorical terms, like a dead organism to corresponding stimulus. Therefore, Western science cannot claim superiority over indigenous knowledge because it cannot be understood as pure representations of the universe seeing that the learner cannot connect with Western science from an experiential perspective. Instead what Western science represents is valuable knowledge, but only as 'temporal notions' in which a learner must re-negotiate his or her understanding or view of the world. This can potentially lead to cognitive conflict that can impede the development of a learner's view of the world that he or she can recall based on experience. This conflict and the way that the learner resolves it are discussed next.

Aikenhead's 'Border Crossing Theory'

Phelan, Davidson, and Cao (1991), who studied students' movements between the micro-culture of families, peers, school and classrooms, found that they struggle to adapt to this new setting without the assistance of people from any other setting. Similarly this enculturation and adoption of science concepts, Aikenhead (1996) asserts, means that the learner must repudiate his or her own worldview in order to embrace the scientific worldview. He argues that it is important to differentiate between enculturation or adoption and assimilation of scientific knowledge. 'Enculturation' is the process by which a learner accommodates scientific knowledge into his world, while making sense of or navigating through the information in order to make sense of it in his or her mental schemata. Assimilation is the process by which the learner subsumes the scientific view and abandons his cultural perspective. Culture refers to the norms, beliefs, values, expectations and conventional actions of the cultural group. In other words, the learners break away from their cultural perception of the world in favour of the scientific viewpoint; they are said to undergo a process called cognitive border crossing.

According to Aikenhead and Jegede (1999), 'border crossing' was first introduced by Giroux in his book entitled Border Crossing: Cultural workers and the politics of education, published in 1992. Aikenhead took the idea of 'cultural border crossing' in science from the work of Lugones (1987) and Costa (1995), who investigated how learners navigate the sense-making process of operating between different family values. Their findings revealed that at times this transition is smooth when the two worlds of family are congruent; at times it is manageable when the two cultures are somewhat different; it is hazardous when the cultures are diverse, but impossible when the cultures are discordant. From Costa, Aikenhead realised that success in science depends on the degree of cultural difference between their life-world and that of school science. In other words, the process of crossing the border from a cultural world to that of science is referred to as cultural border crossing.

Aikenhead (1996) identified four types of border crossing. The first type of cognitive border crossing is referred to as 'smooth border crossing'. This happens when the learner can hold both worldviews at arms length from each other. This means that both the scientific understanding and the cultural viewpoint can exist in harmony in the mind of the learner at the same time. Because these two worldviews exist parallel to each other, the learner can draw from either worldview (i.e. the scientific or the cultural) whenever it is necessary and beneficial to do so. The second type is called 'managed border crossing'. Managed border crossing represents a cognitive shift in the schemata in which the learner's cultural worldview is antagonistic to his scientific worldview, and requires a shift from one worldview to the next that is in need of being managed. The third type of border crossing is called 'hazardous border crossing'. This takes place when the learner diffuses the scientific worldview and the cultural worldview leading to a hazardous transition from one worldview to the next. The fourth type of border crossing is called 'impossible border crossing'. This type of border crossing takes place when the learner completely rejects the scientific knowledge, because it is in direct conflict with the learner's worldview. To such learners their cultural lens dominates their perception of the world and hence they resist the transition to the scientific conception.

Thus cognitive border crossing explains the conflicting worldviews or 'duality of thought' that exist in the mind of the learner and how he or she resolves this tension by either abandoning the one and assimilating the other or he or she can hold two conflicting worldviews at the same time, as a way of dealing with the conflict. Thus, border crossing theory draws attention to the significance of the learner's cultural capital with which he or she from an indigenous setting enters the science classroom and the value of these worldview presuppositions in the teaching and learning process. Border crossing theory also highlights that teachers must become more consciously aware of this cultural capital in the planning of their lessons by using it as a hook, so to speak, to connect the new knowledge or scientific knowledge to when he or she teaches. Next I discuss Jegede's (1999) collateral learning theory to explain how a learner resolves the cognitive tension between Western science and indigenous knowledge.

Jegede's 'Collateral Learning Theory'

Because of the resilience of a learner's cultural framework, Jegede (1995) argues that when a learner is introduced to school science, a duality of thought in their memory and schemata arises. He points out that this is a way how learners cope in a hostile environment. Jegede (1999) identifies four different types of collateral learning that take place in the science classroom when a learner is confronted with Western science. These are parallel, simultaneous, dependent and secure collateral learning. Le Grange (2007) states that these four different types of collateral learning should not be viewed as individualistic or existing in isolation from one another, but must be viewed on a continuum over time as one type can lead to the next.

According to Jegede (1999), parallel collateral learning takes place when the learner holds two conflicting or opposing ideas side by side at the same time. This means the learner stores both the scientific knowledge and the cultural knowledge in his or her long-term memory with minimal interference and interaction at the same time. This is evident, Le Grange (2007) argues, in learners who are introduced to school science for the first time. They assimilate the newly introduced scientific notions and allow them to co-exist with their cultural or traditional worldviews. The second is dependent collateral learning. This process takes place over a longer period of time than parallel collateral learning. This is because by the time learners are introduced to the new scientific information, they might still be processing a concept they learned at home or in their culture. This means the learner has to resolve the tension between the two concepts at the same time, seeing that both are new to his or her cosmos (Jegede, 1999). The third type is dependent collateral learning. This takes place when schemata from one worldview are challenged by those of another worldview, enabling the learner to restructure or modify his or her existing schemata. The shift to the new schemata is triggered by what the learner already knows. This means the learner can make a shift in favour of or against the cultural worldview, depending on what already exists in his or her mind. The fourth and last type of collateral learning is secured collateral learning. Jegede (1999) argues that this type of collateral learning takes place over a period of time in which the learner must resolve the tension between the two opposing knowledge frameworks. This means that the learner must resolve what he termed 'cognitive dissonance' in the knowledge base of the embedded long-term memory. This cognitive dissonance represents the conflicting worldviews of the Western science the learner is confronted with in the science classroom and the cultural knowledge he or she brings into the classroom. To resolve the conflict, Jegede (1999, p. 135) argues, the learner draws from a 'convergence towards commonality' which secures the newly acculturated conception.

Jegede's (1999) collateral learning theory does not explain how the learner experiences each of these types of learning, nor does he indicate a mechanism for how long the learner remains within the ambit of a particular collateral position before moving to the next type of collateral learning. Fakudze (2004) points out that both Aikenhead (1996) and Jegede (1999) agree on the commonalities between their theories of how learners resolve cognitive conflict and that these two theories are interrelated. She states that effective collateral learning in the science classroom will depend on how well the learners can cross the border from the cultural worldview to the scientific worldview. Next I discuss some teaching strategies that science teachers can use in the teaching and learning of science to give due recognition to both Western and indigenous science.

Suggestions for Teachers on How to Teach Science

These theories discussed in the previous sections have shown that a learner's cultural worldview provides the cognitive lens through which he or she engages with the world. According to Cobern (1993), any person's worldview is related to their cultural upbringing, which forms the fundamental organisation of how the mind functions and operates. Hence science teachers cannot turn a blind eye to the worldview presupposition or knowledge that a learner brings into the classroom. This has led to the adoption of a constructivist approach to the teaching and learning of science. The constructivist takes into consideration the environment in which learners live, their sociocultural upbringing, their prior knowledge and so forth in the planning of lessons. The constructivist believes that new knowledge is constructed on the basis of already existing knowledge. The challenge for many teachers is being able to strike a balance between the knowledge a learner brings into the classroom and the new knowledge he or she is required to know at that level. This means that a teacher's pedagogical sequence should be rooted in a bottom-up approach. The 'bottom' represents the prior knowledge (based on personal and cultural engagement with the world), while the 'up' part represents the new content to be taught. In other words, a teacher is expected to design activities that draw from the learner's mental schemata and to link it to knowledge, skills and values that the teacher wants them to learn. From this perspective learning takes place on the basis of already existing knowledge. This raises the question of how a teacher should teach science as a way of acknowledging both worldviews which will be the focus of the next section.

The Teacher's Role as a Cultural Broker

Aikenhead and Jegede (1999) suggest the different roles a teacher can assume in a culturally diverse classroom. They suggest that at times a teacher is expected to take on the role of a *cultural broker tour guide* and at other times a *cultural broker travel agent*. When cultural border crossing is difficult for the learners, the teacher is expected to assume the role of a cultural broker tour guide in which he or she must negotiate meaning on behalf of the learners by taking them to what Le

Grange (2007) calls 'the principal sites in the culture of science and coaches them on what to look for and how to use it [knowledge] in their everyday lives' (p. 588). This means the teacher must exercise and apply various methodical repertoires to get the learner to where he wants him or her to be. For example, in Physical Science electricity is part of the syllabus for Grades 10–12. To explain this phenomenon teachers often draw from environmental illustrations such as lightning to illustrate the concept of static electricity. In the Western tradition lighting is explained scientifically to take place when lighter sub-atomic particles in a cloud move towards the upper (top) part of the clouds which in the process becomes positively charged. Heavier sub-atomic particles in the clouds, on the other hand, move towards the bottom and become negatively charged. When the positive and negative charges become large enough lightning is released between these regions. Most of the lightning takes place between the clouds but some is discharged and strikes the earth in bold flashes. These bold flashes are the discharge of electrons which can kill or injure animals or humans on the ground because it contains thousands of volts that is released in a fraction of a second. In the traditional cultures in South Africa lightning is viewed by many learners from different traditional cultures such as the IsiXhosa's, Zulu's, Sotho's, Venda's and many others as a form of witchcraft. These cultures believe that a 'Sangoma' (traditional healer in their culture) has the power to control and manipulate the environmental forces by making lightning. They believe that Sangomas create lightning to hurt or kill a person or an animal. These two explanations of lightning from the Western and indigenous perspective require the teacher to adopt the role of a cultural broker tour guide in which the teacher must present the material in such a way to minimise cognitive conflict in order for hazardous border crossing and secure collateral learning to take place smoothly.

In situations where the learners need less guidance the teacher can take on the role of a cultural broker travel agent. In this capacity or role the teacher only provides the learners with scientific capital in the form of assignments, tasks, practical investigations, or events from which the learners can develop their own understanding of science. By means of these activities learners acquire the necessary empirical evidence through which they actively engage with science on a personal level which can result in refuting some of their false beliefs and cultural conceptions. For example, in the Western tradition when learners are required to explain the concept of 'power' it can apply to the product of voltage and current (which is measured in Watts or it can also refer to the ratio of work and time . In the indigenous tradition the word 'power' is synonymous with the IsiXhosa word 'amandla' which has multiple meanings such as 'power', 'strength' and 'energy', which has the potential to create confusion in the science classroom.

Scaffolding the learner though smooth border crossings and parallel collateral learning, as well as managed border crossing and simultaneous collateral learning would require that the teacher take on the role of a tour guide as explained in the example of the concept of 'power', whereas hazardous border crossing and dependent collateral learning and impossible border crossing and possibly dependent collateral learning expect the teacher to assume the role of a travel agent as explained in the example of 'lightning'. In all these situations it is imperative for the teacher to draw on the lived world experiences of the learners so that they can relate to the content they are taught in a tangible and meaningful way. This will create an engaging classroom environment in which both the teacher and the learner are actively involved in the sense-making process. If this is the case or the pedagogical approach adopted, the process of teaching and learning will make more sense to learners, seeing that the teacher draws from their lived world experience and it will become more probable that they will be able to accommodate the new knowledge offered to them on the grounds that they themselves understand it. Freudenthal (1991) writes that 'we say we see science as a human activity and that, consequently, science teaching should guide students in *"scientificalising"* their world, instead of trying to transfer scientific knowledge as a ready-made product'. He alludes to the important role of teachers in guiding their learners in a pedagogy of engagement and nurturing them to participate in the knowledge-construction process. This links up well with Bhabha's 'third space' or 'in-between' space for science teaching which is what I discuss next.

Bhabha's Third or 'In-Between' Space for Teaching

According to Bhabha's (1994) notion of the 'third or in-between space for teaching', represented by Figure 2.1, the 'first space' represents the cultural worldview presupposition with which the learner enters into the science classroom. This means the knowledge the learner holds, from the traditional perspective, comes from the meaning and identity given to him or her through the community's indigenous cultural identity. The 'second space' represents the Western science to which the learner is introduced in the science classroom. The intersection between the first and second space (illustrated by the shaded area in Figure 2.1) is referred to as the 'third space'. This space brings together the cultural worldview of the first space and the Western or modern empirical science of the 'second space' to negotiate new meanings and understandings from the two different knowledge practices and language paradigms to resolve the contention between the two worldviews.

Figure 2.1: The model of third space

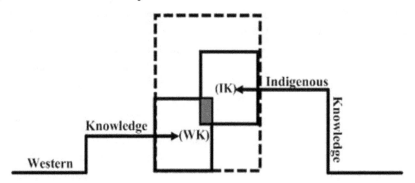

Adapted from: *The Location of Culture*, Bhabha (1994).

Wallace (2004) describes this third space as:

> …an abstraction of a space/time location in which neither the speaker's meaning nor the listener's meaning is the 'correct' meaning, but in which the meaning of the utterance is hopeful for either co-construction of interpretation or new hybrid meanings. (p. 908)

What Wallace points out in the above citation is that when the teacher focuses on this 'third space', he or she moves away from the privileged authoritative discourse and dominance of school science and provides indigenous cultures with improved access to Westernised science, while at the same time embracing and validating local people's own ways of understanding. He argues further and contends that in the 'third space' learners, teachers and communities collaborate to negotiate new meanings that dissolve the dichotomy of 'dual thought'.

However, one of the shortcomings of this model is that it ignores the personal lived world of the learner and the knowledge generated from his or her own personal engagement with the world. According to Merleau-Ponty's (1962) lived body theory, embodiment is a representation of knowledge that children hold to get a grip or put differently to 'cope' with their everyday realities. These realities learners bring into the science classroom subsequently unfold into a third space, namely embodied knowledge. The intersection between indigenous knowledge (IK) (first space), Western knowledge (WS) (second space) and embodied knowledge (EK) (third space) leads to a fourth space (illustrated by the shaded area in Figure 2.2). While in Bhabha's (1994) model the third space is a hybrid space that combines both Western and indigenous elements of culture, I believe it is the fourth space that teachers should target when they teach science, given the overwhelming effect that lived experience has on a learner's outlook on the world.

Figure 2.2: The model of fourth space

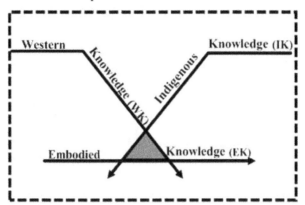

Source: Author.

This raises the question: What materials are required for teachers to effectively direct their teaching aimed at targeting the fourth space?

Firstly, what is needed is for the establishment of new research communities and teachers that teach in the community and that can work with the indigenous people and the learners from the respective communities. Through inter-epistemological dialogue with the locals new materials and coursework that is plausible and relevant to the learners from those communities can be developed. For example, in 2005 in an attempt to help the people of Malawi, the United Nations International Children's Emergency Fund (UNICEF) developed a plan to hand out mosquito nets to ameliorate the spread of malaria (Munyai, 2017). Instead of using the nets to cover themselves when sleeping the Malawians used it for fishing. UNICEF believed that the most important need of the locals was to control the spread of malaria whereas their most urgent need was for basic sustenance. UNICEF was not aware that most African communities sprinkle the urine of cows around the parameters of their houses to fight off mosquitoes that can prevent the spread of malaria. In other words, to avoid such misunderstanding, the main idea is to document the everyday experiences and realities of the local people. Through inter-epistemological dialogue with indigenous people and learners these new research communities can develop a more robust understanding of the perception and knowledge held by local communities. The information that emerges from these dialogues and discussions can be used as a basis for the development of materials and science coursework.

When teachers have a sound understanding of the basic needs and cultural understanding of the communities in which they teach, they can come up with teaching approaches and strategies that focus on the fourth space in their class-rooms. By doing so, cultural border crossing and collateral learning become a two-way flow and no longer a unidirectional pathway of science teaching. This strategy has the potential to overcome one of the main challenges that make learners unre-sponsive to the learning of science, that is, the abstract complex language and ter-minology of Western science that learners struggle to connect with. Acknowledg-ing and merging embodied knowledge and cultural worldviews of learners with Western terminology could result in smooth border crossing. Another outcome that could result from the inter-epistemological dialogue is the development of a more learner-friendly language in which science can be presented. This means that researchers and local communities will have to collaborate in order to develop an indigenous 'vocabulary' for science in the local languages. An indigenous vocab-ulary for science is one way to resolve the tension between indigenous knowledge and Western knowledge. By translating Western science into local languages cre-ate multiple discourses and forms of dialogue that will promote more interest in the scientific description of nature. Next I turn the focus to a phenomenological approach to the teaching and learning of science as an alternative method to meth-odise the delivery of scientific knowledge.

The Phenomenological Instructional Method

The pre-colonial tradition of teaching in Africa was based on the child's active engagement with the world (see Assie'Lumumba, 2012). Consequently most of a child's learning was based on experience that was either modelled by or through practical personal engagement with the physical world. This method or teaching strategy has a lot in common with phenomenological principles both as a method and teaching philosophy. As a philosophy, phenomenology is concerned with the science of lived experience which advocates that without experience there can be no knowledge or effective learning. Conversely, factual knowledge without ex-perience only becomes 'memory' learning with which the child cannot relate at a cognitive level. This is because the perceptual information or mental residues in the memory carry information of which the child has no experience. On the other hand, in a phenomenological paradigm the child's experiences with the real world are automatically linked to a particular sensory modality that releases the information from the long-term memory into the present moment. This is why Husserl (1975) refers to the concept of intentionality, which means that exposure to a particular event leads to an outcome of knowledge by acquaintance. From this angle or perspective it is fair to assume that when a learner is introduced to any information or concept in science, this information is funnelled into two possible streams: an *unresponsive stream* or a *responsive stream.*

The unresponsive stream is activated when the learner has no knowledge about a concept and therefore cannot retrieve any information via the feedback loop from his long- or short-term memory. This unresponsive stream creates a challenge for the child, that is, as the child evaluates any new knowledge, the new information of which he or she has no record cannot therefore make any link in his mind. This can be hazardous for the child's learning as there is a possibility that he or she can reject the information or idea cognitively, as the child does not have any experience to which he or she can relate the information or check its validity and reliability. Drawing from Aikenhead and Jegede (1996) and Jegede (1999), this learning experience could be described as commensurate with hazardous border crossing and secured collateral learning. Because if the information or new knowledge transferred to the child is in conflict with his or her belief or value system, it is only through experience that this tension or conflict can be resolved. This means that 'effective' learning might only take place if the new knowledge can be connected to something tangible that the child can relate to.

The responsive stream is activated when information is imparted to the learner that he or she can relate to. This means the learner already has some kind of experience about the phenomenon and can access/retrieve information from his short- or long-term memory. This makes it easier for teachers to connect the scientific information with the personal experience that the learner brings into the science classroom. This sensory stream can be stimulated further through insightful questioning that could lead the learner towards a deeper understanding of the scientific phenomenon. Drawing on a learner's lived experience automatically activates the responsive stream, which leads to a deeper understanding of the phenomenon as the learner becomes more deeply aware of his or her surroundings. Science can deepen learners' insight into their immediate environment through their close interaction with it. It is therefore important for teachers to use the learners' experiences with the real world as the starting point for every lesson. The following example from Koopman (2017) illustrates how teachers can use lived experiences as a starting point for a science lesson on chemical bonding.

Creating a context for the content by drawing from the learner's personal lived experience:

> Koopman (2017) points out that instead of starting the lesson with an emphasis on the division between the four different bonding types (microscopic level) or the rules associated with each bond type the teacher can create a context in which he or she draws on *an experience,* in the Deweyan sense, in which the teacher relates the scenario to the experiences as lived by the learners.

At this point the teacher can start the dialogue with guiding questions such as: What types of bonds are formed when Mpho blow-dried Tanya's hair? To answer this question the teacher can draw on the burning smell given off in the blow-drying process. This question will set the scene for more layered questions leading to deeper levels of understanding of the topic. For example, what is the nature and classification of disulphide bonds? Are they covalent or ionic bonds, etc.? The teacher can give the learners a task to formulate the differences between the types of bonds. This can be followed through with a discussion on the remark Mpho made about blowing Tanya's hair a little longer to gain an insight into bond strength and bond length, which is temperature dependent, as experienced when her hair shrinking. From that point onwards the learners can be given a task in which they can compare the properties of covalent bonds as depicted in the scenario with ionic bonds as a starting point to unpack the science in the event relating to the different types of chemical bonding in nature. This approach will shift the focus of the lesson from the macroscopic event to the microscopic properties implicit in the scenario. From this point onwards the teacher can guide the discussion to introduce phenomena such as what methods traditional cultures use as a substitute for 'blow drying'. This can lead to a discussion on the question: What type of hair products are used to soften ethnic hair and how these products relate to chemical bonding, bond strength and bond length and their uniqueness and relevance to the different bonds types. This discussion can further lead to modern scientifically formulated products that were developed that applies to all hair types, which can then be related to scientific advances and real-life applications of chemical bonding.

When teachers draw from the personal lived experiences of learners, and indigenous knowledge they automatically open up the responsive stream when introducing science because, if the stimulus (experience) falls outside the receptor field, the 'neurons' short-circuit, leading to meaninglessness or pointless memorisation by learners, so to speak. The explanations of the responsive and unresponsive streams of how knowledge travels through the various mechanisms of the body are intended to urge teachers to expose learners at an early stage to investigate phenomena of interest by designing simple but thoughtful experiments to conjure up or refute hypotheses to expand the child's awareness in the real world. In so doing, teachers would release the learners' passion for asking questions and thus become co-inquirers with them. Teachers could also use the burning questions that learners bring into the classroom as a platform to develop lesson plans and classroom activities to stimulate debate about phenomena that are critical to them. These approaches have the potential to stimulate a deep interest in scientific knowledge per se—and not merely *about* science in general—as learners learn to reflect on their experiences. Therefore, instead of focusing on scientific correctness, learners should be allowed to free their imaginations and raise their awareness about the universe. It is only once learners become aware of what is happening around them that their senses can transport them to new places and explanations about science based on how they feel and experience it. In so doing, they become connected to the universe much in the same way as the great scientific thinkers once experienced the forces of nature through their senses and made new contributions to the world of science.

Conclusion

The research on which this study was based was an investigation to determine how science teachers in South Africa could nurture African learners' understanding of science through effective instructional and pedagogical practices, given the competing/complementary descriptions of Physical Science in the CAPS. To this end, I explained why the rich cultural knowledge that African learners bring into the classroom should be embraced by science teachers. Instead of focusing only on the abstract and incomprehensible Western canonical science teaching, teachers can use the learner's *a priori* indigenous knowledge as a cognitive hook to attach the new knowledge. I also explained how the Western worldview is an impediment to the learning of school science and the possible cognitive dangers it could have for the learning of science. For this purpose I drew from Aikenhead's (1996) border crossing theory and Jegede's (1999) collateral learning theory to articulate and elucidate how this conflict takes place in the minds of the learner and how this tension is resolved.

Finally, I outlined possible approaches on how to teach science to pedagogically engage African learners who struggle with the assimilation of school science. Here I used Jegede's (1999) suggestions of the different roles that a science teacher can take on in the classroom in order to prevent cognitive conflict. Bhabha's (1994) 'third space' was also discussed and extended to a fourth space as another strategy that science teachers need to take into consideration when they teach the subject. Furthermore, teachers are also introduced to phenomenology as an instructional method to assist learners who might experience cognitive dissonance when introduced to school science. How learners respond to school science is an area of study that needs further investigation in order to develop a more robust perspective on cross-cultural science and its effect on learners' cognitive schemata.

References

Aikenhead, G. (1996). Science education: Border crossing into the sub-culture of science. *Studies in Science Education, 27*, 1–57.

Aikenhead, G. S. (2008). Objectivity: The opiate of the academic. *Cultural Studies of Science Education, 3*(3), 581–585.

Aikenhead, G. S., & Jegede, O. J. (1999). Cross-cultural science education: A cognitive explanation of cultural phenomenon. *Journal of Research in Science Teaching, 36*(3), 269–287.

Assie'Lumumba, N. T. (2012). Cultural foundations of the idea and practice of the teaching profession in Africa: Indigenous roots, colonial intrusion and post-colonial reality. *Educational Philosophy and Theory, 44*(2), 21–38.

Bhabha, H. K. (1994). *The location of culture*. London: Routledge.

Cobern, W. W. (1993). College students' conceptualization of nature: An interpretive worldview analysis. *Journal of Research in Science Teaching, 30*, 935–951.

Costa, V. B. (1995). When science is 'another world': Relationships between worlds of family, friends, school, and science. *Science Education, 79*, 313–333.

Department of Basic Education. (2011). *Curriculum and assessment policy statement for the further education and training phase for Physical Science*. Pretoria: Author.

Department of Education. (2002). *Revised national curriculum statement for Physical Science*. Pretoria: Author.

Fadkudze, C. (2004). Learning of science concepts within a traditional socio-cultural environment. *South African Journal of Education, 24*(4), 270–277.

Freudenthal, H. (1991). *Revisiting mathematics education*. Dordrecht: Kluwer.

Hall, S. (1989). Ethnicity: Identity and difference. *Radical America, 23*(4), 9–20.

Hodson, D. (1993). Re-thinking old ways: Towards a more critical approach to practical work in school science. *Studies in Science Education, 22*, 85–142.

Husserl, E. (1975). *The Paris Lectures* (P. Koestenbaum, Trans.). The Hague: Martinus Nijhof.

Jaspers, K. (1997). *Psychopathology* (J. Hoenig, Trans.). Baltimore: Johns Hopkins University Press.

Jegede, O. (1989). Towards a philosophical basis for science education of the 1990s: An African view point. In D. Herget (Ed.), *The history and philosophy of science in science teaching* (pp. 185–198). Tallahassee: Florida State University.

Jegede, O. J. (1995). Collateral learning and the eco-cultural paradigm in science and mathematics education in Africa. *Studies in Science Education, 25*, 97–137.

Jegede, O. (1999). Science education in nonwestern cultures: Towards a theory of collateral learning. In L. Semali & J Kincheloe (Eds.), *What is indigenous knowledge? Voices from the academy (119–142)*. New York: Falmer Press.

Jegede, O. J., & Okebukola, P. A. (1991). The relationship between African traditional cosmology and students' acquisition of a science process skill. *International Journal of Research in Science Education, 13*, 37–47.

Koopman, O. (2013). *Teachers' experiences of implementing the Further Education and Training (FET) science curriculum.* Unpublished doctoral dissertation. Stellenbosch University, Stellenbosch.

Koopman, O. (2017). *Science education and curriculum in South Africa.* New York, NY: Palgrave.

Koopman, O., Le Grange, L., & de Mink, K. (2016). A narration of a physical science teachers' experiences in implementing a new curriculum. *Education as Change, 20*(1), 149–171.

Krüger, R. A. (2008). The significance of the concepts 'elemental' and 'fundamental' in didactic theory and practise. *Journal of Curriculum Studies, 40*(2), 215–250.

Le Grange, L. (2004). (South) African (a) philosophy of education: A reply to Higgs and Parker. *Journal of Education, 34*, 142–154.

Le Grange, L. (2007). Integrating western and indigenous knowledge systems: The basis for effective science education in South Africa. *International Review of Education, 53*, 577–591.

Le Grange, L. (2008). The didactics tradition in South Africa: A reply to Krüger. *Journal of Curriculum Studies, 40*(3), 399–407

Loving, C. C. (1995). Comment on 'Multiculturalism, universalism, and science education'. *Science Education, 79*(3), 341–348.

Lugones, M. (1987). Playfullness, "world"-travelling, and loving perception. *Hypatia, 2*(2), 3–19.

McKinley, E. (2005). Locating the global: Culture, language science education for indigenous students. *International Journal of Science Education, 27*(2), 227–241.

Merleau-Ponty, M. (1962). *Phenomenology of perception* (C. Smith, Trans.). London: Routledge.

Mji, A., & Makgato, M. (2006). Factors associated with high school learners' poor performance: A spotlight on mathematics and physical science. *South African Journal of Education, 26*, 253–266.

Munyai, K. (2017). How learning empathy can help build better community projects for Africa. *Conversation Africa.* Retrieved from https://theconversation.com/how-learning-empathy-can-help-build-better-community-projects-in-africa-75900

Naidoo, P., & Lewin, K. M. (1998). Policy and planning of Physical Science education in South Africa: Myths and realities. *Journal of Research in Science Teaching, 35*(7), 729–744.

Ogunniyi, M. (1987). Conceptions of traditional cosmological ideas among literate and non-literate Nigerians. *Journal of Research in Science Teaching, 24*(2), 107–117.

Ogunniyi, M. B. (1988). Adapting Western science to traditional African culture. *International Journal of Science Education, 10*, 1–9.

Ogunniyi, M. B. (2007). Teachers' stances and practical argument regarding a science—Indigenous knowledge systems: Part 1. *International Journal of Science Education, 29*(8), 963–986.

Phelan, P., Davidson, A., & Cao, H. (1991). Students' multiple worlds: Negotiating the boundaries of family, peer, and school cultures. *Anthropology and Education Quarterly, 22*, 224–250.

Seltzer-Kelly, D. (2013). Feynman diagrams, problem spaces and the Kuhnian revolution to come in teacher education. *Educational Theory, 63*(20), 133–149.

Shizha, E., & Emeagwali, G. (2016). *African indigenous knowledge: Journeys into the past and present.* Rotterdam: Sense Publishers.

Wallace, C. S. (2004). Framing new research in science literacy and language use: Authenticity, multiple discourses, and the 'third space'. *Science Education, 88*, 901–914.

CHAPTER THREE

The Dangers of Poor Science Teaching

Introduction

Several South African studies show that Science teachers frequently encourage rote learning (Muwanga-Zake, 2008; Nganu, 1991; Rogan & Grayson, 2003; Rollnick, Allie, Buffler, Campbell, & Lubben, 2004). Consequently, their learners rely heavily on memorisation to pass examinations. According to Chisholm and Wildeman (2013, p. 90), this approach to teaching comes as no surprise, since there were already high-stakes examinations as early as 1915. The authors further point out that, after the transition to democracy in 1994, South Africa adopted a 'post-bureaucratic model' of accountability in which the quality of teaching was measured by learner performance as a strategic plan to monitor teachers (Chisholm & Wildeman, 2013). The emphasis placed on learner performance as a measure of teacher effectiveness encourages teachers to neglect the foundational knowledge that learners require to master science and results mainly in teachers adopting teaching methods of teaching to the test (*ibid.*). This, they argue, undermines the purpose of education, because by teaching to the test leaves learners underprepared and with no confidence or competence in science. Teaching to the test to a large extent forces teachers to select and teach pre-designed dogmatic canonical science. This brings me to the aim of this study, which is to elucidate the dangers of such 'poor science teaching'.

To this end, this study chronicles the lived experiences of four experienced Physical Science teachers in exploring the meaning and implications of 'poor' science teaching. 'Poor' science teaching is defined as teaching styles that encourage memorisation of facts and rote learning; fail to link science to real-life experiences; are textbook bound; and deliver abstract and obscure knowledge with little or no regard for practical work. This study is an attempt to provide a conceptual framework on what goes on in the minds of these teachers when they teach the content to their learners. To this purpose, insights into the long-term consequences of being exposed to 'poor science teaching' themselves assisted to explain how their thinking was decoded or recoded as learners under apartheid education. Furthermore, I hope to illustrate the tensions that continue to exist in their minds as a result of the ways they were taught and their struggle to break free from that. The insights gleaned from this investigation are important, as the findings might go beyond common sense explanations and metaphors to account for why teachers struggle to change their pedagogical practices. This study also intends to raise awareness amongst policy-makers, curriculum planners and prospective Physical Science teachers of the dangers of 'poor' science teaching, its origins and how the subsequent psychological tensions can control a teacher's pedagogy and decision making for decades to come.

Rationale of the Study

The aim of this study is to chronicle the educational experiences of four 'experienced' Physical Science teachers in South Africa. It explores how the teaching methods of their own Physical Science teachers and lecturers impacted on their understanding of science, which potentially discursively shaped their teaching practices as teachers later in their lives. Furthermore, an attempt is also made to understand their past, as learners, and how their teachers and lecturers established a pre-existent identity of science teaching through their pedagogies. This study shows that in order to develop an understanding of why teachers do what they do when they teach, it is important to look into the totality of their lived experiences rather than discrete and isolated moments. For this purpose it is important to note that all four teachers who participated in this study received their schooling and their university training within the framework of Fundamental Pedagogics (FP; explained in detail later).

Most studies in science education in South Africa over the last three decades focus on curriculum issues, such as the nature of the knowledge and its relevance to South African learners (Le Grange, 2008; Ogunniyi, 2000; Rogan & Grayson, 2003), while others focus on learner and teacher content knowledge, teachers' pedagogical knowledge, teachers' disposition towards the teaching and learning process. But no research (as far as my reading of the literature indicates) has delved into the consciousness of Physical Science teachers or, more specifically, *what* Physical Science teachers think about when they teach and *why* they think in that particular way. The term 'thinking' refers broadly to the way they view science, how it should be conveyed to their learners and why they view it in that particular way. International studies affirm that the professional 'self' is deeply rooted in beliefs and values which emanates from how they were thought and trained as teachers (Talbert, 1995). However, this study takes the concept of 'self' or 'teacher identity' further by focussing on the object of Physical Science teachers' consciousness.

Indeed, the search for answers to these kinds of questions seems particularly apt for phenomenological research, as the findings could be viewed as the centre of conscious experience. Certainly, shifting the focus to exploring human consciousness through questions such as 'what Physical Science teachers think about' and 'why they think in that particular way' might assist in ameliorating some of the major challenges plaguing the effective teaching of Physical Science. This has the potential to provide insight into the development of adequate and appropriate professional development programmes for teachers, how to train prospective teachers in order to make the subject more meaningful, appropriate and relevant to their learners, and to assist curriculum planners and advisors in curriculum design and development.

By researching lived experiences we could gain an insight into the subconscious mind of teachers revealed through a voice that speaks directly to us. This will unveil their everyday realities and allow us to see and understand what they feel when they teach and how they cope in the complex world of teaching. It can also give us direct access to their anger, frustration and fears because of the pressure placed on them by society, the media and government to produce good results. We might learn from them that teaching Physical Science is not simply about relaying/delivering knowledge to learners, but also about the survival of the soul and body, and the struggle to break free from the suffocating dogmatism of the way that they were trained as learners and students. These insights will allow us to see their struggle to be open to change in the face of new ideas and innovative ways of teaching. This study resonates with the lived world experiences of many Physical Science teachers in South Africa who were exposed to similar teaching approaches. For example, empirical studies done by Naidoo and Lewin (1998), Ogunniyi (2002), Muwanga-Zakes (2008), Le Grange (2016) and Koopman, Le Grange, and de Mink (2016) substantiate this claim and report that the struggle for effective science teaching in South Africa has been a long one that continues to this day. A consequence of this long struggle has given Physical Science a high-priority status since the transition from apartheid to post-apartheid South Africa. The justification for elevating science to a high-priority subject is because the nation's economic growth and development is directly related to the effectiveness of its Physical Science programmes. In addition, the demands created by globalisation with the rise of the 'knowledge economy' require the need for more scientists such as engineers, medical scientists, chemists, astrophysicists, and so forth. Therefore, this study is necessary as it shifts the focus from the tensions and conditions external to the teachers themselves that they have to engage with on a daily basis towards what happens in the minds of these teachers when they teach. This focus on the internal worlds of the teachers allows us to see how their past education both as learners and teachers of Physical Science discursively shaped their identities as teachers.

Next I provide the political context of the study to explain the education system and its philosophy within which the teachers were trained as learners and teachers before they entered the schools as Physical Science teachers.

Political Context

A Brief Overview of Fundamental Pedagogics (pre-1994)

Seeing that all four teachers in this study received their schooling and university training within the framework of the hugely influential, all-pervasive educational doctrine of FP, which was closely aligned to the ideology of Christian National Education (CNE) at the time, this section gives the reader a brief overview of how FP developed in South Africa and how it was applied as a doctrine to produce an inferior educational experience for black learners. This brief background will help to shed light on the essentials of FP, such as its ideological underpinnings, its autonomous approach to the teaching and learning environment, and the interrelationship between the teacher and learner in the delivery of the curriculum content.

In the mid-1970s and early 1980s FP was the official doctrine that determined the nature of instruction and learning in all schools, Afrikaans-medium universities and historically black colleges and universities. According to Le Grange (2008), the doctrine of FP, which was built on the ideology of CNE, emerged in 1944 as an instrument through which apartheid education was later 'legitimised'. Enslin (1984) states that CNE was promulgated as a policy for white, especially Afrikaner, learners, as part of the rationale for educating black people for unequal participation in economic and social life. She states that, according to CNE policy for black education, instruction should (i) be in the mother tongue, (ii) not be funded at the expense of white education, (iv) preserve the 'cultural identity' of the black community and (v) be administered by whites only. Articles 14 and 15 of the CNE policy specifically state that 'black education is the responsibility of white South Africa or more specifically of the Boer nation as the senior trustee of the native...' (Enslin, 1984, p. 401).

Naidoo and Lewin (1998, p. 732) point out that these South African education policies separated schools within the Republic of South Africa and those within the so-called self-governing territories of Venda, Ciskei, Transkei and Bophuthatswana (which were later reincorporated into the 'new South Africa'). Within the boundaries of the Republic of South Africa, the education system was administered through four main departments: the House of Assembly (DET-A) for those classified as white, the House of Delegates (DET-D) for Indian people, the House of Representatives (DET-R) for coloured people and the Department of Education and Training (DET) for black people. In addition, some provinces were further sub-divided into their own educational systems.

During this period the proportion of unqualified Physical Science teachers employed in schools classified under the Group Areas Act of 1966 was as follows: coloured 43%, black (Africans) 87%, Indian 4% and white 2% (Naidoo & Lewin, 1998). The Physical Science curriculum under apartheid rule was delivered to learners in an excessively theoretical and old-fashioned way (Nganu, 1991). According to Kahn (1993, p. 8), this was as a result of a teacher-centred regime that was strict, inflexible and dominated by examinations. The syllabi were outdated. At a conference held in Johannesburg in September 1993 the ANC expressed the view that the Science curriculum was trapped in a traditional and abstract paradigm, with very little attention being given to the everyday experiences of learners. Too much emphasis was placed on facts and rote learning. The end result, according to the Foundation of Research and Development (1993), was that, out of the entire school learner population in South Africa, 47% of white learners choose Physical Science, compared to 14% of Africans, with the average pass rate for white learners being as high as 90% and for black learners lower than 10%.

Krüger (2008) traces FP back to the work of Oberholzer (1955, 1968), Nel (1968), Landman, van Zyl, and Roos (1975), Van der Stoep (1969) and various others, who were all members of the education faculty at the University of Pretoria. The first FP publication in South Africa, *Inleiding tot die Prinsipiële Opvoedkunde* [Introduction to Fundamental Pedagogics] by Oberholzer (1955) led to FP becoming a powerful doctrine in the 1960s, 1970s and 1980s. According to Oberholzer (1968), FP has relatively autonomous accounts of knowledge that is pre-designed and that discloses the structures of phenomena (p. 170). Consequently, the important task of FP is for the teacher to describe the phenomenon so accurately that there is no basis for debate or argument (Landman cited in Barnard, 1992). The pedagogical environment does not provide room for deeper engagement with and exploration of the meaning of the phenomenon, nor should it encourage any open-minded dialogue, discussion, or exploration of ideas in science. Furthermore, it does not view the subject content as an opportunity to train learners to live responsibly. For example, in contemporary South Africa the Curriculum and Assessment Policy Statement (CAPS), which is underpinned by a constructivist philosophy, envisages a learner who is open-minded, critical and capable of using content to save lives. This means that when the instructor teaches Charles' Law in the chemistry classroom, learners are expected to see the link between the law (that an increase in pressure causes an increase in temperature) and accidents on the road. In other words, they must not only know the law, but must be able to apply it effectively in everyday life.

Instead, in FP science was taught in a way that promoted only the 'fundamentals' of the content, while the broader details and applications of the content were ignored. According to Le Grange (2008), the fundamentals had a strong theoretical focus and formed the essential framework within which the teacher approached the learning environment and the design of classroom activities. In the process the teacher was viewed as the one who unlocked reality for the child in the form of statements without discussion or further deliberation. Such an FP approach compartmentalised knowledge and promoted a behaviourist philosophy. In this system, with its strong emphasis on content, the child was perceived as an empty vessel waiting to be filled (i.e. the learner was the passive recipient of knowledge).

Although there was such a strong emphasis on content, this approach did not adequately prepare learners for careers in science. For example, under the mechanics section (which included vectors and scalars, forces, momentum and electricity) the focus was only on rectilinear motion (one dimension), which was very elementary. At university, the focus of these topics was on two- and three-dimensional motion. This shift from one to two and three dimensions confused learners because it required spatial perceptions and scientific equations that were different from those which they had been taught in school. Furthermore, solving such problems required more complicated mathematics. This dumbing-down of standards was described as deliberate, because motion in two dimensions formed part of the curriculum in the same standards/grades on other parts of the African continent.

The doctrine of FP promoted a behaviourist theory of learning, which will be discussed next.

How Fundamental Pedagogics was Implemented

The idea of FP, according to Kruger (2008, p. 216), was initially underpinned by a phenomenological philosophical framework for learning that makes the learner and his lived world the centre of the teaching and learning environment. Kruger writes 'I was immersed in phenomenology…and saw the person ("man") as existential and open to meaning-given possibility' (p. 217). In this paradigm the focus is not on the learning of facts but in discovering the ultimate truths about the universe through personal lived experience. For this reason Kruger concludes FP must be grounded in 'understanding the fundamentals' of human existence, which must place, in no specific order, the 'reality', 'lived moments', *'dasein'*, 'embodiment' and so forth as the beginning and end of all learning. Thus FP was believed to be grounded in Heidegger's human science concept of *'dasein'* which means *'to be'* or *'being there'* (Kruger, 2008). Kruger drew on the work of Heidegger (1967) to explain what is meant by *dasein* in the context of FP. He sees *dasein* as a person's epistemic nature and views the learner's existence not as cognitive activity, but as concrete in the sense that it involves real beings who are physically present in the world and whose knowledge of the world is derived from their personal interaction with 'concrete physical things' in their immediate life world (p. 58). These visible and concrete physical things can only be experienced through the senses. Although FP in theory was built on these wonderful promises of a purely phenomenological didactical approach, what materialised instead was an abstract, empty, imaginary and cognitive world of words and concepts at the classroom interface (see Le Grange, 2008). This happened because un- and under-qualified teachers had to deliver the content to learners (Naidoo & Lewin, 1998). Therefore instead of adopting a phenomenological approach to teaching, the teachers adopted a behaviouristic approach to learning.

The theory of behaviourism was first postulated in 1925 by the psychologist John Watson in his book entitled *Behaviourism* (Rabil, 1967). Watson's theory, as noted by Rabil (1967), originated from a critique of empirical psychology in which he gave Ivan Pavlov's data greater significance. To Pavlov psychological functioning could be explained from observed behavioural data. Such observable data led to a breakthrough and the proposal of a scientific methodical structure to predict how organisms behave and subsequently learn. The term behaviourism covers three separate doctrines, namely (i) metaphysical behaviour (i.e. there is no such thing as consciousness; there are only organisms behaving); (ii) methodological behaviourism (i.e. true psychology can only study publicly observable behaviour and does not deal with introspection) and (iii) analytical behaviourism (i.e. psychological concepts can be analysed exclusively in behavioural terms).

The behaviourist theory of development views the mind as a machine. The machine (e.g. an optical scanner) registers (sensory) experiences in the individual's receptive mind much in the same way that a teacher consigns knowledge to the learner's memory using different educational media such as an epidiascope or overhead projector. Alternatively, this theory can be likened to the way a computer stores bits of information (the process of data input) and later retrieves them when needed (process of data output or retrieval). An example at this point might help to understand how the behaviourist theory of learning was implemented. If two learners are exposed to the same pictures and images of the planet Jupiter and two days later they are asked to imagine the planet Jupiter, the chances are that both learners will have the same thoughts. This is because both learners have no personal experience or physical connection with the planet Jupiter. Apart from conjuring up images of Jupiter in their minds, the exercise does not require them to use their senses and therefore they cannot think further than the information which the teacher has given. What both learners' thought patterns have in common is not empirical and causal content, but ideal theoretical content independent of the senses. It is reasonable to infer that any follow-up question relating to Jupiter will be beyond their reach and this can be described as a 'woeful blindness' with respect to what a real image of Jupiter looks like, or why one needs to know anything about Jupiter. This example relates to the way in which FP was used as a powerful tool to exercise control over the learner that aimed to exclude the use of the senses in the meaning-making process.

Underlying the behaviourist metaphor of the mind as a machine are the associationist or stimulus-response theory, which views both the specific and general cognitive structures as reflections of structures that exist outside the learner's world. Cognitive development is the result of guided teaching and learning. The efficacy of such theories is measured and limited by specific desired outcomes. For example, during examinations or tests, a learner's performance is measured only by how well he or she can recite or regurgitate information from textbooks and class notes.

On the basis of Heidegger's (1967) philosophy of perceptual experience, through the *dasein* the lived world experiences of the learner should have been embraced; instead it was ignored and most lessons encouraged rote learning and memorisation. In other words, *dasein* was not applied with reference to its original meaning and intent and subsequently learners were not encouraged and guided to use their senses to classify objects according to their colour, shape, structure, mass, length and so forth, an approach which lies at the heart of phenomenology. Moreover, learners were not introduced to different objects, challenges, stimuli or making observations through which they could learn to construct their own ideas about how they think objects might behave in the surrounding real world or in a controlled space. Consequently they were unable to make scientific predictions that allowed them to generate data to formulate or falsify hypotheses to construct their own rule book of science. This is because the objective of FP, under the false pretence of adopting phenomenology as an approach to learning, was to control mindsets and to narrow the learner's horizon in line with political pressures. This eventually led to a content-heavy and examination-driven approach to teaching which consequently did not have the learners' interests at heart. Thus FP which obscurely reduced phenomenology to an examination-driven approach discourages rather than encourages learners to start asking deeper questions, which might prompt them to conduct further investigations. Furthermore, teachers became more textbook-bound and little, if any, attention was paid to practical investigations.

Research Design

Drawing on the work of Husserl and Heidegger, this chapter uses a phenomenological data-construction process to reveal the natural attitude of each research participant that is deeply embedded in his or her consciousness. What makes phenomenology different from other methods is that it does not offer the possibility of a theory with which researchers can explain the world of participants; instead it offers the possibility of describing the lived world of the participant with a high degree of accuracy (Van Manen, 2007). This means that the phenomenologist brings a particular kind of sensitivity to the field of educational research, something which, according to Van Manen (1990), has been long overdue. I thus had to be receptive and avoid the dangers of being misled, side-tracked or enchanted by extraneous elements, and avoid getting carried away by unreflective preconceptions and personal emotions. So instead of theorising about the participants, I focused on the significance of what they were saying as they described the lived experiences of their teachers and university lecturers and their own practices as Physical Science teachers.

Sampling

Four Physical Science teachers were purposively selected. Cook and Campbell (1979) note that purposive sampling is a useful way of collecting data, if it is carefully applied. They argue that the use of purposive sampling depends on the nature of the research question under investigation. Datallo (2010), on the other hand, asserts that purposive sampling is used, amongst other things, to select a small subsample in order to closely examine typical and unusual or extreme elements. From this perspective an array of factors, such as race, qualifications, age, gender, region and teaching experience, was considered in the selection process. Table 3.1 gives a succinct overview of the participants who were selected.

Table 3.1: Overview of the research participants

Participants	Race	Age	Years of experience	Gender	Qualifications	Major subjects
A	*Xhosa*	*46*	*19*	*Male*	*Diploma in Education*	*Physical Science Mathematics*
B	*Coloured*	*56*	*25*	*Male*	*BSc, BSc (Hons), HDE*	*Botany, Zoology*
C	*Coloured*	*65*	*5*	*Male*	*BSc, HDE, BEd (Hons), MEd*	*Chemistry, Botany*
D	*Coloured*	*63*	*27*	*Male*	*BSc, HDE*	*Botany, Zoology*

Source: Author.

All the participants received their schooling under the National Assembly Training and Education Department 550 curriculum before 1994. All of them are males, of whom three teach in the Northern District and one teaches in the Eastern District of Cape Town in the Western Cape Province. Respondents B, C and D, who teach in the Northern District, labelled themselves with the historical term 'coloured', and both C and D have more than 25 years experience, while C has only 5 years experience of teaching Physical Science. This is because respondent C worked for 25 years as a laboratory technician in the oil industry, before enrolling for a teaching qualification at the age of 59 years. All of them were Afrikaans mother-tongue speakers and all attended poorly resourced schools as learners. Furthermore, B holds an Honours degree in Botany as a major subject, C a Master's qualification in Science education, while D has an undergraduate degree and a teacher's diploma in Education. Participant A teaches in the Eastern District and referred to himself as African. According to the Department of Basic Education's criteria for qualifying as a Physical Science teacher, none of them qualifies as none of them did Physics at least to second-year level. A is also not qualified to teach Physical Science at FET level, as he has only a diploma in Education. He is an IsiXhosa mother-tongue speaker with 19 years teaching experience. The Northern District of Cape Town is rated one of the most successful districts in the Western Cape Province based on teacher commitment and learner performance in the National Senior Certificate (NSC) examination. For example, in the last five years the Northern District produced outstanding Grade 12 results in the Western Cape. Most schools situated in the Eastern District, on the other hand, struggle to produce such consistent results in the NSC examination.

Interviews

Two in-depth, open-ended, unstructured interviews with each participant were used to collect the data for the study. This method provided each research participant with the opportunity to tell their stories about how they were trained as learners, what they think about when they teach and why. According to Price (2003), artful interviewing takes place when the researcher knows and understands the ways in which people's thoughts, beliefs and actions correspond with each other. As such, the laddered technique of questioning as advocated by Price (2003) was used to elicit rich descriptions about their lived world. Laddered questions operate on three levels, namely (i) inviting descriptions aimed at setting the scene and making the respondent feel that the researcher is interested in what he/she has to say or offer; (ii) knowledgeable or invasive questions are asked later in the interview, when the respondents have shown signs of relaxation or comfort. This involves questions such as: What do you think? How do you feel? and (iii) questions of personal philosophy. These are the most invasive questions. These questions focus on beliefs, values and deep-seated feelings. This is the core to the respondent's personal identity. Asking questions at this level is akin to asking questions about who you are and may leave the respondent feeling that the researcher is judging them. To apply the laddered technique effectively I decided to divide the interview into two parts.

The first part of the interview, which was roughly one hour long, focused only on inviting questions, such as where they were born and raised, their childhood experiences, where they attended high school and the environment in which they had to learn Physical Science, who their Physical Science teachers and lecturers were, how they (teachers and lecturers) taught the subject and what their views were about his or her teaching. These questions allowed me to get to know them (although on a superficial level). To win their trust, I showed an interest in their responses by listening carefully to what they had to say. I also showed empathy through my body language and tone when I responded to their answers and descriptions of their lived world. The answers to these questions provided an insight into their childhood experiences, including the resources of their schools, their views about their teachers, the teaching methods of their high school teachers and lecturers and whether they were adequately prepared for careers in science or teaching at the institutions for higher learning at which they studied. The data provided me with an overall perspective on how they learned science and whether any consideration was given to practical work. It also assisted in understanding how they were pre-discursively shaped and influenced for careers in science.

The second part of the interview started off with knowledgeable questions, such as how they teach the subject, and what they consider to be important when they teach. For example, do they value practical work, do they focus more on tests and examinations. Furthermore, I focused on whether they transformed their own practices from the way they were taught science at high school or university. I then shifted to more invasive questions that invited emotional responses by asking questions such as why they became Physical Science teachers, what motivates them to stay in the profession, how they cope with the challenges, what they think about when they plan and teach their lessons. The answers to these questions allowed me entry into their personal philosophies and deep-seated emotions about the subject.

Throughout each interview I paid careful attention to each participant's embodied selfhood to get a sense of 'who they are' in order to explain the ways in which the 'self' is constructed. In other words, I explored how they see themselves emerged in the world of teaching, or as Heidegger puts it, how they cope being-in-the-world of the Physical Science classroom. At all times I attempted to abstain from imposing any preconceived or *a priori* views of the world by placing myself in their shoes in order to view the world through their own experiences without any form of bias.

Next I discuss the data explication framework.

Data Explicitation Framework

Constructing the Descriptive Narrative Using Husserlian Phenomenology

I applied Husserl's famous dictum, that is, to return to the things themselves (*zu den Sachenselbst*) (Husserl, 1975, p. xix). In other words I did not attempt to make assertions about that which I could not see in the data; *Sachen*, which refers to subconsciously held ideas, were the focus of my analysis. My aim was to 'report' and 'explicate' the structural nexus that gives their experiences as learners specific meaning, depth and richness. The structural nexus refers to the patterns or units of meaning that form the unifying whole in each participant's contextual setting through a process of reflection. I was not only interested in how they learned science, but more on the essence or nature of their experience from their point of view (Van Manen, 1990). This provided deeper insight into the effect of a particular learning experience. Lived experience does not need any interpretation because, according to Husserl (1970), the interpretation already exists *in the experience.*

Husserl used the phrase *epoche* to explain the conduct of bracketing when reading a transcript. I consequently entered a space free of any pre-suppositions and suspended all possible interpretations and preconceived meanings. This required me to iteratively read each respondent's transcript with openness and enter the individual's world to extract meaning from what the person is saying. It must be stated that at times this was difficult, because each participant has his or her unique way of experiencing temporality, spatiality and materiality, but each of these coordinates was understood both in relation to others and to their own inner world (Hycner, 1985, p. 29).

Constructing the Interpretive Narrative Using Heideggerian Phenomenology

Heidegger (1967) believed that consciousness is not separate from the world and is context dependent. Heidegger's philosophy allowed me to unpack the context in which the teachers grew up, the schools they attended and the teaching strategies they were exposed to as learners and at university, His conceptualisation of being (*dasein*) allowed me to engage with the transcript of each participant in terms of how they comprehended and perceived the content as learners, how they internalised the teaching methods of their teachers and university lecturers and how they projected these experiences in their own personal practices as Physical Science teachers.

Heidegger's (1967, p. 3) notion of *dasein* allowed me to link their present with their past to develop a framework to understand why they do what they do when they teach Physical Science.

An Overview of Their Transcript Analysis

I applied both Husserlian and Heideggerian phenomenology to each transcript as follows:

- After iteratively reading each transcript I highlighted key words in every transcript as they related to the foundational experience and meaning given to the main research question and themes;
- I collated these key terms after removing all extraneous, unrelated and superfluous items to derive the central themes in each participant's transcript. This allowed me to gain insight into each participant's phenomenological attitude. From this I could write the descriptive narrative by focusing on the things themselves;
- By combining the sub-narratives from the net of data, I constructed the interpretive themes;
- The interpretive themes were constructed by highlighting high-frequency words and the priority given to certain words in their responses. I selected two interpretive themes. These themes were augmented with my own personal experiences as an ex-teacher of physical science and adumbrate some of the literature related to personal and professional sense of self;
- I then crafted my own descriptions of what I thought reflected the inner consciousness of the participants as expressed from their perspective.

Findings

The data presented below focused on the following three aspects as they emerged from the transcripts:

- childhood experiences of each participant teacher;
- participants' views of their Physical Science teachers and how they perceived their teachers' teaching approaches and
- how participants teach and perceive science given the context in which they were taught at school and university.

Table 3.2: Biographical data of the participants

P	Descriptive biography	Researcher's notes
A	I grew up in Khayelitsha. I attended school in the township with no resources. I am 46 years old. I am the first person in our family to study at a university. I completed my three-year teaching diploma at a Technikon in Physical Science. It was my dream to become an engineer one day, but my marks was too low, so I became a teacher. I had to accept being a teacher because now I can change how learners learn science because my teacher clearly failed me. I have 19 years teaching experience.	IsiXhosa first language user. Grew up in abject poverty. Attended a 'township' school with poor resources. He believes his teacher destroyed his childhood dream of becoming an engineer. Has 19 years teaching experience and chose teaching as a second career choice. Completed a three-year diploma at a Technikon.
B	I grew up in Bishop Lavis. I am 56 years old. I grew up very poor and went through many struggles. I wanted to be a plant pathologist so I majored in Botany and Zoology at UWC where I studied. I could not complete my degree because of financial problems and began to teach in my second year. I did some physical science in my second year so I could teach the subject. I completed my degree five years later and thereafter did a Hons degree in Botany. I did this while I was teaching but took a year off. I have 25 years experience	Afrikaans first language user. Grew up in a poor environment. Studied towards a degree and left after two years to teach. He finished his degree five years later. Well qualified to teach Life Science, as he has a lot of experience with Botany and Zoology. He can offer more to learners in Life Science. He has 25 years experience. Childhood dream was to become a plant pathologist.

P	Descriptive biography	Researcher's notes
C	I grew up in the rural town called Beaufort-West. I'm 65 years old. I did not do PS at school as our school did not offer it. I fell in love with science when I saw how my father fixed broken radios. He was the local handy man in our town. I am teaching now for five years. Before this I was in the oil industry for 25 years, the Chemical Engineering sector, where I actually used the knowledge gained at University. I majored in Chemistry…and did some Physics. I studied at UWC. I could attend UWC due to apartheid. I finished off my degree in 1972, in the days where it was not good for any coloured person getting a BSc degree to go to industry because it was accepted that you should become a teacher. I believe I was not born to be a teacher so I wanted to go to Industry, but look where I am now, I am a teacher	Afrikaans mother tongue user. Grew up in a rural town. His love for science was inspired by his father. He enrolled for a degree in Chemistry with no school science experience. He worked for 25 years in the oil industry. Well qualified to teach PS.
D	I am 63 years old. I grew up in Pietermaritzburg in Durban. As a child we lived in a very dry and arid part of Durban. My parents are teachers but did not have a lot of money due to the political situation. But at least I attended a good school and we had a lot of resources. I studied at UWC and majored in Botany and Zoology. I did Physics and Maths at university in my 1st year. Although I did not want to become a teacher I had to do so because it was the only career for me. Otherwise I had to become a policeman and that I did not want to do.	English mother tongue user. Both his parents were teachers. Attended a good school. Completed a degree in Botany and Zoology. He did not want to be a teacher. Political situation in the country forced him into teaching. More qualified to teach Life Science and not PS as he majored in Botany and Zoology.

Source: Author.

Table 3.3: A description of the teaching styles of the participants' Physical Science teachers and lecturers

Descriptive phenomenology	Interpretive narrative
A Looking back today, I hate my science teacher and what he did to us. I don't want even remember this guy's name. In Grade 12 we had to teach each other the work. He was absent a lot and when we reported him to the principal nothing happened. He was not even qualified. No practical work was done. I did not understand the work. He read the content word for word from the textbook that is how he taught us. I will never forget this guy. It's because of him that I could not become an engineer. My marks were too low, so I opted for teaching.	Inadequately prepared for a career in science and science teaching. He had an unqualified PS teacher who taught poorly which he hated. Was taught science as rote memorisation of facts and meaningless data (Reddy, 2006). No practical work which can obscure the relevance and meaning of science (Dube & Lubben, 2011).
At Technikon not much has changed. Although we learned a lot of new things and did some practical work, our lecturer used the same way of teaching as my teacher. Some things we had to learn on our own and we wrote a lot of tests so that we can know the content when we teach.	At Technikon this 'poor' teaching continued since the focus was on knowing the content as opposed to understanding it. Tests and examinations were the focus of the lessons.
B Despite our poverties and difficulties I still made the choice to move to another school to do PS. I found myself in a school where the PS educator was unqualified. His teaching method was the entire class had to read through the textbook from one page to wherever the bell took us. That actually was his method for three years so we had no teaching of PS—we actually became very good readers to the advantage of our Afrikaans and English educators. We never had a problem with our reading, each one of us never knowing that we read every day for one full period one paragraph of text book.	He shows resentment and deep-seated anger towards his PS teacher. He had a 'poor' school science experience. Inadequately prepared for careers in science and teaching. No higher order critical thinking skills as they had to read from the textbook. Forced to become independent. He showed frustration for the theoretical discourse that dominated his learning. His lecturer followed a 'chalk and talk' approach. Again he had to take responsibility for his own learning since the focus was on examinations. Rote listing of facts instead of practical science was the order of the day. No context-driven and antithetical relation to his lifeworld (see Onwu & Stoffels, 2005).
At university our practicals and theory classes were split. There again the lecturer talks for hours and they don't care if you listen or understand. They don't even care if you are there because we were too many. But the time is short and they have to do a lot of work. So it is examination driven and here we had to do a lot of self-study. They did not prepare us well for the world of work.	

Descriptive phenomenology	Interpretive narrative

C Looking back at all the sad stories I heard of PS teachers I think I'm one of the lucky people that did not do Physical Science as a school subject in the old Standard 10. I did general science till Grade 8. I'm thinking my liking for Science came in because my father himself was very scientifically inclined. He believed that whatever man has made, you can tackle; whatever is broken and it can be repaired if you got the right equipment. He was very innovative in the sense that you should improvise, you must try out new things, you should explore. He himself had no qualification but yet he went to study how to repair radios and television so, that equipped him to become sort of a handy man in the town where we lived. Well, while I was in Matric I realized the need for knowing more about Physics and Chemistry, and on my own I studied through a correspondence school which was called Success College that time. I studied through them and I in fact studied Physics and Chemistry; however, the sadness of the apartheid system was such that we could not write more subjects than what the school offered. The so-called coloured people took 6 basic subjects and you could not write another subject outside of that. So I never got so far as in fact writing PS as a Matric subject; however, I believe that the knowledge gained then was sufficient to help me with the challenge of university.

I studied pure Chemistry at university and it was a lot of work. I had to do a lot of self-study. We did a lot of tutorials, tests and every term examination. You can imagine I had to learn these formulas by heart and when we got class he just talk and talk. Sometimes I did not even hear what he said and just wrote down notes and studied it. But we did a lot of practical work. At least one practical a week and a test. So we had to know how to use the equipment.

No formal school training of PS. Learned only the practicality of science with little understanding of the processes. As a child he could see science at work and its application to real-life phenomena. His quest for understanding the processes of science encouraged him to become a scientist. Watching his father searching for a solution to a problem (when he fixed stuff) transformed his life. It gave him inward meaning. He knew he wanted to become a scientist. He sees the experiences of others who did PS at school as a crisis, when he says 'I am one of the lucky ones not to do science'. He learned Physics and Chemistry through a correspondence college and had to learn the work on his own.

At university he was exposed to poor science teaching where he had to learn most content on his own. Lecturer followed a 'chalk and talk' approach. Focus of the teaching was on knowing facts which had to be memorised. On the positive side they did a lot of practical work which prepared him well for industry and for teaching. This implies he has higher-order critical thinking skills which he learned from experience in industry. Bennett and Holman (2002) state that practical science helps to demonstrate the relationship between theory and practice.

Descriptive phenomenology	*Interpretive narrative*
D It was factual information, there was not really in any way a link with what you did in class and what is physically happening outside. Occasionally you talk to other people and you discover they went through the same kind of thing. Poor teaching. The textbook was like the Bible. We had to know the facts like Bible rhymes. I think back sometimes, and realise there were times when you had enough general knowledge to make sense of the work. So you could relate to things that you've heard and discovered and read about and it made sense quite honestly most of the time. I don't think my teacher understood the work he was teaching us, because I could not see the link with real life. My lecturers at university only gave us a prescribed textbook to buy. Here we had to underline the main topics and take notes at the same time. Most of the time I only wrote the notes down and studied it like a parrot because you can't listen and write. But it was just 'chalk and talk' for 3 more years. I was so used to it that all I wanted was to finish my degree and work to earn money.	He describes his teaching style as 'poor teaching'. Teacher used the textbook as his bible. Textbook orientated teaching. Looking back, he thinks his teacher did not understand the content because the work was not related to meaningful events outside the classroom. When this happens, Onwu and Kyle (2011) state, teachers fail in their task to make learners see the inward reward and potential personal benefits of science. At university he had to learn the work from his textbook and did a lot of self-study. His lecturer used 'chalk and talk' method and he had to rely on memory learning. Memory learning is meaningless or shallow learning (Reddy, 2006).

Source: Author.

Table 3.4: Narrative data on their own practices as teachers

P	**Descriptive narrative**	**Interpretive narrative**
A	Most of the time I have to work through question papers to prepare learners for the examination. We don't have a lot of time for practical work. The year is short and time is not always on our side. So we have to try to even come in over weekends to teach our learners. It is tough, but what can you do. It is all about marks here at our school. I feel guilty for not doing what I am supposed to do, because you don't do justice to the subject.	He follows the same 'poor' teaching approach as his teacher. Constandi (2000) states we teach who we are and how we were taught. He teachers to the test and focus only on questions papers. Drawing from Foucault (1978) he becomes an agent of government. He does not think practical work is important therefore he focus on theory. He uses the examination to escape the guilt he has towards his learners.

P	Descriptive narrative	Interpretive narrative
B	The passion to do Physical Science practically is there, and that has been the hallmark of my teaching up till now. You [researcher] are two weeks late, …two weeks ago you would have noticed that this school was used to train Grade 12 PS educators in this area in practical work. As a matter of fact the dirty stuff is still right at the back of the room that the subject advisor used to do the training. This school has been instrumental in teaching Physical Science educators in this whole north district in practical work since 2003, whenever there is practical work that need to be taught to educators, practical workshops is done here. But I think we must shift the focus to train learners, by now teachers should be able to do this on their own. But I struggle to find the time because the department pushes us for the marks. In fact they have now scaled the number of practicals per year down so that we can do more teaching.	He wants to adopt a more practical approach to teaching science and considers it the 'hallmark' of his teaching career. Consciously he has a particular mode-of-being (*dasein*) in the science classroom but does not act accordingly (Heidegger, 1967). *Dasein* has disposition but the context of needing to produce good marks hampers his becoming. He feels the department must place more emphasis on practical science as opposed to rote learning. However, he feels the time is too limited for practical work as he must teach more content for his learners to pass the examination (synonymous with Foucault's (1978) concept of bio-power).
C	After finishing my HDE I taught PS to Grade 11 and 12 learners. I have such a lot of information to share with them because of my experience in the oil industry. We did a lot of practical work and activities but I later realized this not going to work as I had to prepare them for the examinations so I focus on what they need to know to pass the examination. It is all about pass averages here and not how much information you can teach your learners. But I try to stimulate their appetite to learn science practically, although we don't always have the time to do so.	He claims he offers science in a practical way but restricted by examinations (Foucault's complexity of power is at work from top-down (state) and also from the bottom-up (teacher). Although he considers practical science important, consciously he is constrained to teaching the content. Therefore he switched his interest to rote and memory learning. He occasionally then did practical work.

P	Descriptive narrative	Interpretive narrative
D	It can become very difficult to teach effectively in a school with no resources because as you know to make the subject practical requires time and money. We don't have that luxury. In fact it's gotten worse now since we shifted to these new curriculums. So I focus mainly on what the learners must know to pass the examination. I try mostly to do practical work that links topics with activities so that they can see the link between theory and practice. But I don't always have the time as they start to write exam early. I try to keep my eyes on the examination programme and plan accordingly.	He teaches in a school with limited resources. He focuses mainly on factual knowledge in his teaching and what the learners need to know to pass the examination. He does practical work on selected topics to consolidate the content. Consequently, he cannot make his learners to see the personal value and rewards of science.

Source: Author.

Discussion

Bio-Power as an Emerging Strand

The findings reveal a particular mode of being shaped by issues of 'power' and 'control' circulating from the top-down (government) and from the bottom-up (poor teaching). Firstly, power is exercised over each participant teacher's mind because the findings reveal overlapping characteristics such as 'teaching to the test', examination-driven approaches, lack of practical work and rote or memory learning. There appears to be a close correlation between 'who they are' as Physical Science teachers and 'the way they were trained or taught as learners' and students. Secondly, power represses who they want to be. For example, they all want to be teachers that value the world of the learner by adopting constructivist pedagogies. However, the findings reveal that they are constrained by so much attention given to test and examination results. From this perspective poor teaching has consequences for the future of the subject, that is, how the subject will be offered to future generations. This is corroborated by the way they describe their own practices. For example, Teacher A states: 'I use exam question papers when I teach'. Teacher B points out: 'I struggle to find time [to do practical work] because the department pushes us for marks'. Teacher C wants to make a difference by focusing on meaningful learning but 'focus(es) on what his learners need to know for examinations', while Teacher D states that he keeps his eye on examinations and then plans accordingly. Foucault's (1979) functionality of power or the 'limit experience' explains how poor teaching forces a particular mode of being that limits a teacher. Poor teaching was deeply entrenched in their minds—so 'hard wired' in their brains—that it forces 'trans-personal action', that is, they will affect their learners negatively in their own practices as teachers. To paraphrase Masschelein (2006) (who draws from Foucault) the findings revealed that the teachers' 'limited experience' as learners or 'ex-position' historically constitutive of their consciousness guides their practice. This 'limit-experience' as proposed by Foucault does not refer to a body of knowledge, a determinate set of ideas or competencies but the struggle to break away from the old selves.

According to Apple (1979), schools are consciously set up and run in the belief that they support and maintain state aims and ideologies. Consequently, schools support the contours and configurations of a government's agenda for the learner, leading to what Foucault (1978) refers to as subjectification of the learner. Thus the learning 'experiences' of children in reference to Foucault can be read as prominent ways in which government discursively shapes and control the mindsets through learning spaces. Apple (1979) argues that power and culture should not be seen as separate discourses, but as attributes that shape a mode of being. Thus, the notion of 'mode of being' triggers a very important question, namely, who are they [the teachers]? More precisely, what did they become? Their exposure to poor teaching gives us a glimpse of who they are which can be read as a form of 'governmentality'. Foucault (1996) writes:

> What one seeks then is not to know what is true or false, justified or not justified, real or illusory, scientific or ideological, legitimate or abusive. One seeks to know what are the ties, what are the connections that can be marked between mechanisms of coercion and elements of knowledge,…. (Foucault, 1996, p. 393)

In this citation Foucault alludes to the connection between chosen elements of knowledge, power and its subsequent effect on our existence. The exposition of FP gives us insight into how each research participant's mindset regarding science was shaped. Their experiences in the science classroom can be described as a 'limit experience' which restricted their understanding and outlook of science and world which is now repeated in their own practices as teachers. The ties and connections as a result of mechanisms of coercion are clearly visible in the practices of all four teachers (A, B, C, and D). Firstly, they were all exposed to poor science teaching. Secondly the way they were taught as learners reflects the way they teach the subject. There is compelling evidence in the literature that one teaches who one is (Constandi, 2000). All of them teach to the test and in the process give little consideration to the meaning-making processes of science. Thirdly, they see innovative teaching strategies and practical science as less important than knowing facts, concepts and definitions for the examinations. The findings revealed that their epistemic dispositions towards science was shaped and influenced by their own teachers and lecturers.

Next I shift the discussion to the images of their thoughts and why they think the way they do.

Images of Their Thoughts

To explain the images of their thoughts the article draws on Heidegger's (1967) concept of *dasein* from two angles: (i) past memories and unconscious complexities, and (ii) entangled regimes of power and knowledge. The findings reveal that their past experiences as learners are embodied as 'technologies of memory' as they adopt the same teaching approaches. As learners they were exposed to painfully frustrating pedagogies, while in their own personal practice the issues of inquiry in science were viewed as matters of method and protocol. The problem regarding their teaching and/or learning of science as method and protocol is not that it is unscientific or surreptitious, but rather the problem is the way in which these discourses pervade their learners' real-world experiences. Also, it is not their methods and procedures per se that are an issue, but rather their struggle to overcome the discourses they learned under FP and to view the teaching and learning of science as a creative and critical practice. Consequently, their past memories of a 'limit-experience' which led to a 'limited understanding' of science as learners played a major role in how they view the subject, which developed into a framework of thinking emanating from a historic-philosophical consciousness that aligns 'past events with future confrontations'. They use this framework to select and organise the content according to what they think is important, legitimise the content from their perspective and align it with state aims and objectives. By so doing, however, they disrupt the learners' intellectual development and understanding of science.

In the teachers' minds exist conflict between old ideas and new ways of teaching that they find difficult to overcome. Consequently they experience a struggle to revert to the kinds of activities that foster critical thinking and learning amongst their learners that will aid their learners to understand their environments, allowing them to break through the boundaries and limits that FP had imposed on their own thinking. It is this stranglehold that imprisons their thoughts and makes it a struggle for them to transform their practices. This perhaps might explain the confusion in their perceptions of how important science is for the learners, but their teaching is restricted to the textbook and examination question papers. They want good pass averages instead of well-informed learners who are adequately prepared for careers in science. For example, Teachers A and D followed exemplar question papers and the textbook religiously to prepare their lessons, while Teachers B and C follow a practical approach, although 'limited' and 'constrained' by time and preparation for the examination. The implications of these approaches suggest indirectly that they find it very difficult to stop their old ways of doing things. This means that they lay claim in their thinking space to the familiar FP approach, which makes it difficult for them to transform their thinking.

All of the teachers agreed that 'practical work requires time and resources'. According to Foucault (1983) power circulates in a top-down way (the state) which subjectifies the 'self' (teachers). Subjectification as a result of a 'limit experience' (of school science) creates a framework of thinking that developed into a philosophy with which they approach the teaching of science. Consequently, through their obscure philosophy they view their personal practices or teaching approaches as a 'praxis of practice'. This so-called 'praxis of practice' becomes an essentialist discourse with which they justify their 'poor' teaching. Foucault points out that perceived meaning as internalised in the conscious mind does not flow from an objectivistic space but from both a social and subjective space. For example, the teaching strategies of their own teachers and university lecturers, the situation in the school and so forth point to these social and subjective spaces. All these aspects are influential and shaped their mindsets. This culminates in a discursive formation which represents how they were socialised and eventually unconsciously internalised these poor teaching experiences that they were exposed to—for example, inadequate understanding of obscure content, no or little practical work and the focus on the textbook, tests and examinations. In other words, they draw from past experiences to make sense of their own world of teaching. What consolidates their mindsets further is what Foucault refers to as the non-discursive influences, such as pressure from the department, pass averages, time constraints and so forth. Therefore, their narrow perception of science as content to be learned in the absence of context is buttressed by a fear of transforming their practices and has its roots in the poor teaching of their Physical Science teachers and university lecturers.

Implications

The findings in this study might resonate with the lived experiences of many Physical Science teachers in South Africa. For this reason it is important to investigate how a teacher's life history shapes (or informs) his practice. The findings revealed the negative impact or damaging effect that their early or past experiences as learners and students of Physical Science might have had on their understanding of the subject. Thus it is fair to ask if these teachers who participated in this study were so negatively affected by poor teaching as learners, chances are that many others might also be affected, because many teachers currently in the system also received similar training as learners and pre-service teachers. Therefore this study could offer new insights into how curriculum policy-makers and curriculum advisors might provide training to teachers. Furthermore, the study might also reveal that Physical Science teaching is a highly complex phenomenon, especially with regard to what it means to be a Physical Science teacher in contemporary South Africa. This is because of the demands that the DBE, society in general and the pressure of the Grade 12 results place on teachers. The study showed the respective challenges that might face teachers on a daily basis as well as indicating their perceptions about the subject. As a result, the study might open up new insights for curriculum policy-makers and training officials on how to structure their planning so as to position to meet the individual needs of teachers more effectively. This study might also inform curriculum advisors on how to orientate themselves in the field for the pedagogic benefit of teachers, bearing in mind the uniqueness of each teacher's experiences. The findings showed that teaching is much more than the dutiful delivery of the curriculum, but also entails the necessity of being sensitive to the teachers' needs in order to ensure that they do the right thing for the learners.

Conclusion

This article has shown that there are strong links between the research participants' life history which shaped their epistemological development, their understanding of science and how they teach the subject. From this (epistemic) perspective, the findings raise an awareness of two salient points: firstly, their understanding of science was narrowed down by a 'limit-experience' rooted in training as learners and students of Physical Science. This 'limit-experience' had a direct bearing on how they taught the subject. Secondly, their learning or educational experiences occurred under unique social and political circumstances which at a very fundamental level explains why they teach the way they do. Therefore poor teaching cannot always be viewed as an act of the will, but at times (like in this study) emanates from exposure to bad experiences as learners. It is fair to assume that the effects of their becoming Physical Science teachers have the potential to spill over to future generations. The findings also showed the devastating impact that poor science teaching can have on learners. This study has helped us to see and think about what is lying in the shadows of the teachers' subconscious mind.

References

Apple, M. (1979). *Ideology and curriculum*. London: Routledge.

Barnard, F. (1992). The significance of philosophy for the student of fundamental pedagogics. *South African Journal of Higher Education, 6*(1), 7–16.

Bennett, J., & Holman, J. (2002). Context-based approaches to the teaching of Chemistry: What are they and what are their effects? In J. K. Gilbert, O. De Jong, R. Justi, D. Treagust, & J. H. Van Driel (Eds.), *Chemical education: Towards research-based practice* (pp. 165–184). Dordrecht: Kluwer.

Chisholm, L., & Wildeman, R. (2013). The politics of testing in South Africa. *Curriculum Studies, 45*(1), 89–100.

Cook, T.D., & Campbell, D.T. (1979). Quasi-experimentation-design and analysis issues for field setting. Boston: Houghton Mifflin Company.

Constandi, S. (2000). Meandering through my epistemological patchwork quilt: A narrative inquiry of my landscape of learning. *Journal of Philosophy and History of Education, 50*, 86–92.

Dattalo, P. (2010). Ethical dilemmas in sampling. *Journal of Social Work Values and Ethics, 7* (1), 1–17.

Dube, T., & Lubben, F. (2011). Swazi teachers' views on the use of cultural knowledge for integrating education for sustainable development into science teaching. *African Journal of Mathematics Science and Technology Education, 15*(3), 68–83.

Enslin, P. (1984). The role of fundamental pedagogics in the formulation of education policy in South Africa. In P. Kallaway (Ed.), *Apartheid and education: The education of black South Africans* (pp. 139–147). Johannesburg: Ravan Press.

Foucault, M. (1978). *The history of sexuality*. New York, NY: Pantheon Books.

Foucault, M. (1979). *Discipline and punish: The birth of the prison*. New York, NY: Vintage Books.

Foucault, M. (1983). *Der Wille zum Wissen. Sexualität und Wahrheit 1*. Frankfurt: Suhrkamp.

Foucault, M. (1996). What is critique? In J. Schmidt (Ed.), *What is enlightenment? eighteenth-century answers and twentieth-century questions*. Berkeley, CA: University of California Press.

Foundation of Research and Development. (1993). *South African science and technology Indicators*. Pretoria: Author.

Heidegger, M. (1967). *Being and time* (J. Macquarrie & E. Robinson, Trans.). London: SCM Press.

Husserl, E. (1970). *The crisis of the European sciences and transcendental phenomenology: An introduction to phenomenological philosophy* (D. Carr, Trans.). Evanston, IL: North-Western University Press.

Husserl, E. (1975). *The Paris Lectures* (P. Koestenbaum, Trans.). The Hague: Martinus Nijhof.

Hycner, R. (1985). Some guidelines for the phenomenological analysis of interview data. *Human Studies, 8*, 279–303.

Kahn, M. (1993). *Building the base: Report on a sector study of science education and mathematics education. A product of great consultation carried out for the Commission of the European Communities.* Johannesburg: Pretoria and Kagiso Trust.

Koopman, O., Le Grange, L., & de Mink, K. (2016). A narration of a Physical Science teachers' experiences of implementing a new curriculum. *Education as Change, 20*(1), 149–171.

Krüger, R. A. (2008). The significance of the concepts 'elemental' and 'fundamental' in didactic theory and practise. *Journal of Curriculum Studies, 40*(2), 215–250.

Landman, W. A., van Zyl, M. E., & Roos, S. G. (1975). *Fundamenteel-pedagogieseessensies: Hulleverskyning, verwerkliking en inhougewig [Fundamental pedagogical essences: Their appearance, actualization and giving them content].* Durban: Butterworth.

Le Grange, L. (2008). The didactics tradition in South Africa: A reply to Richard Krüger. *Journal of Curriculum Studies, 40*(2), 399–407.

Le Grange, L. (2016). *Rethinking learner centred-education: Challenges faced by the African child when learning school science and maths.* Unpublished paper.

Masschelein, J. (2006). Experience and the limits of governmentality. *Educational Philosophy and Theory, 38*(4), 561–576.

Muwanga-Zake, J. W. (2008). *Is science education in a crisis: Some of the problems in South Africa. Science in Africa, Africa's first online magazine.* Retrieved from www.scienceinafrica.co.za/scicrisis.htm.

Naidoo, P., & Lewin, K. M. (1998). Policy and planning of Physical Science education in South Africa: Myths and realities. *Journal of Research in Science Teaching, 35*(7), 729–744.

Nel, B. F. (1968). *Fundamentele orienteering in die psigologiese pedagogiek [Fundamental orientation in psychological pedagogics].* Stellenbosch: University Publishers and Booksellers.

Nganu, M. (1991). *Overview of African countries' strategies in tackling problems of science, technology and mathematics education in human resource development for the post-apartheid South Africa.* London: Commonwealth Secretariat.

Oberholzer, C. K. (1955). Die teoretiese onderbou van en praktiese toepassing uit die Pedagogiek [The theoretical foundation and practical application from the pedagogical] *In Feesbundel aangebied aan Dr Barend Frederik Nel, hoogleraar in die pedagogiek aan die Universiteit Pretoria, by geleentheid van sy vyftigste verjaardag op 16 Desember 1955/Jubilee Album Dedicated to Doctor Barend Fredrik Nel, Professor of Education in the University of Pretoria, on the Occasion of his Fiftieth Birthday on 16 December 1955* (pp 127–142). Pretoria: van Schaik.

Oberholzer, C. K. (1968). *Prolegomena van 'n prinsipiële pedagogiek.* Kaapstad: HAUM.

Ogunniyi, M.B. (2000). Teachers' and pupils' scientific and indigenous knowledge of natural phenomenon. *Journal of Southern African Association of Research in Mathematics, Science and Technology Education, 4*, 70–77.

Ogunniyi, M.B. (2002). Border crossing into school science and the contiguity learning hypothesis. Paper presented at the *NAST* conference.

Onwu, G., & Kyle, C. (2011). Increasing the socio-cultural relevance of science education for sustainable development. *African Journal of Research in Mathematics, Science and Technology Education, 15*(3), 5–26.

Onwu, G., & Stoffels, N. (2005). Instructional functions in large, under-resourced science classes: Perspectives of South African larger classes. *Perspectives in Education, 23*(3), 79–91.

Price, B. (2003). Laddered questions and qualitative data research interviews. *Journal of Advanced Nursing, 37 (3):* 273–281.

Rabil, A. (1967). *Merleau-Ponty: Existentialist of the social world.* New York, NY: Columbia Press.

Reddy, V. (2006). The state of mathematics and science education: Schools are not equal. In S. Buhlungu (Ed.), *State of the nation: South Africa, 2005–2006* (pp. 392–416). Pretoria: HSRC Press.

Rogan, M. R., & Grayson, D. J. (2003). Towards a theory of curriculum implementation with particular reference to science education in developing countries. *International Journal of Science Education, 25*(10), 1171–1204.

Rollnick, M., Allie, A., Buffler, A., Campbell, B., & Lubben, F. (2004). Development and application of a model for students' decision making in laboratory work. *African Journal of Research in Mathematics, Science and Technology Education, 8*(1), 13–27.

Talbert, J. (1995): Boundaries of teachers' professional communities in U.S. high schools: Power and precariousness of the subject department. In L. S. Siskin, & J. W Little (Eds.): *The subjects in question: Departmental organization and the high school.* Teachers College Press: New York.

Van der Stoep, F. (1969). *Didaktiesegrondvorme [Didactic ground forms].* Pretoria: H & R Academia.

Van Manen, M. (1990). *Researching lived experience: Human science for an action sensitive pedagogy.* London: Althouse Press.

Van Manen, M. (2007). Phenomenology of practice. *Phenomenology & Practice, 1,* 11–30.

CHAPTER FOUR

How FET Physical Science Teachers Teach Selected Chemistry Topics

Introduction

The Curriculum and Assessment Policy Statement (CAPS) for the Further Education and Training (FET) band calls for Physical Science teachers to be innovative, whilst valuing 'high knowledge and high skills' (Department of Basic Education [DBE], 2011, p. 4). One of the many interpretations of what innovation means revolves around the notion of increased quality learning, while the notion of high knowledge and skills refers to the setting of high but achievable goals that teachers must set for their learners. There are many reasons why CAPS calls for innovation and high knowledge and skills, but the main reason is to promote effective science teaching. According to Treagust, Chittleborough, and Mamiala (2003) effective science teaching, more specifically effective Chemistry teaching, hinges on how well the teacher can explain abstract and complex chemical concepts and phenomena. Harrison and Treagust (1996) claim that to do this Chemistry teachers must be able to engage with all three levels of representation: the macroscopic (as a concrete, visible phenomenon), the sub-microscopic (with reference to its invisible properties) and the symbolic (by means of equations representing the changes that take place in a chemical reaction and the numerical values of atoms). The aim of this study is to ascertain how this might materialise during a professional development workshop with teachers.

The last two decades have witnessed a growing concern about the low level of subject matter knowledge and pedagogical content knowledge of science teachers (Rollnick, Bennett, Rhemtula, Dharsey, & Ndlovu, 2008; Stott, 2013; Umalusi, 2011). These concerns are based on the assumption that a distorted understanding of science among teachers can impede the efficient practice of science education. For example, Basson and Kriek's (2012) research into the content knowledge of science teachers in South Africa revealed that among 68 science teachers in township schools more than 80% found the content intimidating and above their teaching ability. The researchers concluded that 80% of the teachers in those areas were unqualified or under-qualified to teach Physical Sciences. Studies conducted by Rogan and Grayson (2003) and Stott (2013) found that, because of the poor content knowledge of science teachers and the fact that some teachers are trapped in traditional pedagogies, learners perceive the subject as difficult.

Onwu and Randall (2006) conducted investigations in Nigeria, Japan and South Africa to ascertain students' understanding of the micro and sub-micro properties of substances and their relatedness to the macroscopic properties. The investigation in South Africa with first-year pre-service teachers revealed that the students struggled to make the links between the three levels of representation. They also found that many students had intuitive misconceptions of triplet Chemistry. A study by Ramnarain and Joseph (2012) on triplet Chemistry with Grade 12 learners revealed similarly that learners could understand discrete levels, but experienced difficulty in grasping the interconnectedness across all three levels. Although there has been research on Grade 12s' and first-year pre-service teachers' operationalisation of the three fundamental levels of representation in Chemistry, there is a dearth of empirical studies involving South African teachers' classroom enactment of the interconnectedness of all three levels.

This significant research gap prompted me to embark on this project to resolve the question: How do science teachers engage with all three levels of representation of Chemistry when they teach? Specifically, the research study aimed to determine science teachers' enactment of the macro, sub-micro and symbolic levels of selected Chemistry topics in their classroom teaching. The findings of this study are important in shedding light on why learners may find the subject difficult so that systematic reflections can be made for curriculum development and the training of teachers.

Three Levels of Representation

Scholars from different parts of the world adopt various naming systems for the three levels of representation of objects in Chemistry. For example, Andersson (1986) refers to this as the 'macroscopic' and 'atomic world'; Bodner (1992) uses the terms 'macroscopic', 'molecular' and 'symbolic' Chemistry, whereas Johnstone (1993) calls them the 'macro', 'sub-micro' and 'representational' levels of Chemistry. This study uses the terms *macroscopic, sub-microscopic* and *symbolic* at an individual level and in a triplet relationship to encompass all three levels of representation.

According to Treagust *et al.* (2003) who conducted a similar study to this one in Australia, the macroscopic level of representation points to the observable chemical phenomena that individuals experience through the use of their senses. This involves a wide range of observation skills such as colour changes, alterations in shape when new products are formed and reactants disappearing. Locke (2009) argues that when individuals experience phenomena through their senses (sight, smell, touch, sound, taste), the sensual encounter penetrates deep into their minds, creating several distinct perceptions based on the peculiar properties of sensual contact. The brain retains these observed sensory images and leaves an impression which is stored in our long-term memory. Chemists routinely communicate macroscopic events using symbolic levels that include 'pictorial, algebraic, physical and computational forms such as chemical equations, graphs, reactions mechanisms, analogies and model kits' (Treagust *et al.*, 2003, p. 1354). Gilbert and Treagust (2009) describe microscopic-level entities as involving particles that cannot be seen with the naked eye and require chemists to extend the capacity of their senses by using optical microscopes. To explain these observations in terms of sub-microscopic particles, chemists often build models of entities such as atoms, molecules and free radicals. The material world at its foundational level is made up of sub-microscopic particles and is synonymous with the DNA of objects. This means that the explanation for the internal and external properties of objects—such as shapes, colours, brittleness, tensile strength and toughness—is to be found at the sub-microscopic level. Understanding the microscopic world of objects requires an understanding of the electrical properties of atoms, bonding angles, bonding energies, bond lengths, electron densities, the movements of orbitals and intermolecular forces. Harrison and Treagust (1998) explain that there is a range of misconceptions among students and teachers about the sub-microscopic level because of the particulate nature of matter, the inability of human beings to engage with it physically and their need to rely on imaginary representations.

It follows that a good understanding of Chemistry hinges on how well a teacher can establish the interrelationship between these three levels of representation, because the symbolic and sub-microscopic levels contain the theory that explains the macroscopic representation of objects. As Bucat and Mocerino (2009, p. 12) explain: 'Seeing as a chemist is a fact-supported, theory-laden exercise of a lively imagination.' In other words, good Chemistry teachers give their learners an insight into the deeper meanings behind the facts and allow them to see what chemists see when they observe a phenomenon. The opposite is also true, as science teachers with inadequate content knowledge of the triplet levels of representation can create serious obstacles to their learners' ability to understand Chemistry. Therefore Treagust *et al.* (2003) assert that expert Chemistry teachers (i) present the content in a meaningful way connecting the three levels of representation at a level appropriate for the learners without oversimplification; (ii) use relevant artefacts; (iii) embrace the learners' prior knowledge of phenomena and (iv) challenge the learners. Kozma and Russell (1997) claim that most novices use only one form of representation, fail to convert to other levels and consequently mostly focus only on the macroscopic descriptions. Van Berkel, Pilot, and Bulte (2009) assert that most novice teachers are textbook-bound and reproduce superficial chemical explanations. This is because textbook writers generally represent expert Chemistry knowledge superficially and therefore may create misconceptions in explaining naturally occurring processes and events. Cheng and Gilbert (2009) report that, instead of teachers invoking the skills to connect all three levels of representation in the science classroom, learners have to rely on diagrams, visual aids and pictures in textbooks to make the representational connections with chemical processes such as the movement of electrons during electrolysis, electrochemical cells and in the Haber–Bosch process.

The sub-microscopic level is often considered a complex terrain for teachers because it requires the use of meticulously precise descriptive language. This level of representation, Van Berkel *et al.* (2009) argue, is best left to the learner's imagination, instead of teachers attempting to explain and describe this level, as this could lead to possible misconceptions. When teachers leave the sub-microscopic level of explanation to the learner's imagination, teachers develop and instil open-mindedness and creativity in their learners. This holds true, they argue, because it indirectly encourages learners to become co-inquirers in the process, thereby allowing them space for personal engagement with the sub-microscopic level.

Teachers' Differing Verbal Explanations in Chemistry

In a study in Australian classrooms Treagust *et al.* (2003) developed five different types of frequently used teacher explanations of chemical concepts. These explanations are:

i. *Analogical*, in that teachers use personal real-life experiences to explain unfamiliar concepts and phenomena;

ii. *Anthropomorphic*, as they reduce complex scientific concepts to recognisable human attributes;

iii. *Relational*, because teachers draw on their own personal engagement with objects or events to explain phenomena;

iv. *Problem-based*, because concepts are explained by drawing on hypothetical problems as a basis to connect with the content, which is common in calculations and

v. *Model-based*, as scientific models are used to explain phenomena.

The current study investigates these explanations closely to describe how teachers in the study connected and engaged with the content of Chemistry in the science classroom.

Methodology

The Research Participants

A total of 15 'lead Science teachers', including four female teachers, were selected by the DBE to participate in the study. Lead teachers were chosen with consent in cluster meetings, and are normally more experienced than others. They are considered competent in their subject field and have a record of excellent results. Lead teachers, in their clusters in the respective districts, are often required to share their experiences and expertise with others and mentor novice teachers. Only one of the teachers had limited experience of 2 years, whereas all the others had between 10 and 25 years experience. All the teachers were TshiVenda mother-tongue speakers and represented 15 different districts in the Vhembe area.

Instructional Context

The research team (or facilitators) presenting the workshops were a consortium of academics from three different universities, specialising in Mathematics, Chemistry, Physics, Zoology and Botany, respectively. The author of this article is a Chemistry expert. The research team engaged the teachers in a week-long workshop once a year for two years. In these workshops each facilitator presented at least two topics per day from the Grades 8 to 12 syllabus. In every Chemistry topic all three levels of representation were focused on and the importance and value of the links between them was discussed. Rather than teaching them in these workshops how to deliver the content, participants were encouraged to share their ideas of how they think it best to present the content.

On day three of each workshop the teachers divided themselves into five groups of three teachers. No criteria were given for this division into groups. Each group had to plan and present a lesson on a topic of their choice from the Grade 10 syllabus to their peers (pretending to be Grade 10 learners). The brief given to the teachers was:

- Take one aspect of the syllabus and plan and present a 30-minute lesson;
- Each lesson has to include all three levels of representation in Chemistry which must be integrated into the content;
- The content must be relevant and captivating to the learners;
- The prior knowledge of the learners must be tested;
- The content must be challenging and not oversimplified.

The teachers were given more than enough time to exchange ideas and to construct their lessons. Subsequently the lessons were presented in the conference venue whilst each group member had to make a contribution. During each lesson a video was made of the whole group. After the lesson presentation, a group discussion followed where each lesson was dissected to determine whether or not the presenters had integrated the three levels of representation into the content. The responses were probed where necessary to ensure that there were no misconceptions or misunderstandings. The data sources used for this study were video recordings, field notes and post-lesson interviews.

Data Analysis

The analysis focused on the different levels of representation tracing how each group included, used and engaged with them in their lessons. I first watched all the videos many times and augmented the field notes as necessary to develop an understanding of how the teachers in each lesson integrated the three levels of representation. After analysing the videos and drafting my main findings, I invited a colleague, an expert in Physical Science, who was unfamiliar with the data, to watch the videos with me in order to validate my findings. We looked at the videos together and encapsulated the different levels of representation and discussed my main findings. This led to a refinement in the descriptions of the distinctions between the different levels.

Guided by Treagust *et al.*'s (2003) typology of explanations, we also classified the verbal explanations of the teachers for each lesson and how they described concepts, the accuracy of their descriptions and noted during the analysis their main explanations. All disagreements between us were resolved by consensus.

Findings

Lesson 1: The Atom

The teachers divided the class into three groups. Each group had to use a water bottle and a glass (with a narrow circular base and a larger top part) to draw three concentric circles in their note books. They were then guided by the teachers as follows:

> Group one, I want you to draw three small plus signs inside the first circle, group two six and group three eight.... Now I want each of you to draw 2 small circles on the line of the second circle...eight small circles on the line of the third circle.

They then asked the class the following question: 'From your memory in Grade 8 of electricity, which charge do you associate with a plus?' This was followed by a long pause, after which one group responded, 'it's a positive charge'. The teacher agreed and complimented them, and explained: '...in Chemistry those that are plus are called protons, which means it is one part of an atom that is found inside the nucleus'. The teacher then continued: 'One part is positive and the circles that you drew on the lines outside the inner circle are the electrons. I now want you to follow this up by drawing inside the smaller circles that were drawn on the line with minus sign to indicate the presence of electrons.' They then instructed the respective groups to go to their periodic tables and to identify the atom that they have drawn by name by looking at the number of protons in the nucleus. In the next activity the different groups were instructed to draw an atom with 11 protons, 13 protons, 15 protons and 17 protons, and to identify and name the atom by using their periodic table.

Lesson 1: Interpretation

Throughout the lesson the teachers focused mainly on the sub-microscopic properties of elements. Little connection was made with the *symbolic* properties and no link was made with the *macroscopic* properties of elements. When they were asked in the post-lesson interview why they ignored these two levels they said, 'our focus was on drawing a model of the atom, we thought these concepts were not important to know...'. Applying Treagust *et al.*'s (2003) framework of teacher explanations, it showed that their lesson followed a *model-based* approach. The lesson was based mainly on an imaginary model of the atom. To some extent, their explanations adopted *relational explanations* as they drew on the learners' own personal engagement with electricity to explain what protons are.

Lessons 2: The Periodic Table

The teachers started the lesson with the following narration:

> As far as the periodic table is concerned, I often look at it as a bible. When you go to church, you carry a bible with you, so the preacher uses the bible to teach you about God. I want you to know that, just as the preacher knows the bible, Chemistry uses the periodic table to explain everything. The most important periodic laws are the groups and the periods. Groups are the vertical rows of elements whereas the horizontal rows are referred to as the periods. As indicated here [teacher pointing to the different groups on the periodic table] we have Group 1, 2, up until group 8.... The last ones, in Group number 8, are called inert gases or noble gases. They are also identified as the kings.... They don't mix with the other elements if they don't have to. Therefore all the other elements on the periodic table aspire to be like them.

They continued by focussing on the line that separates the metals from the non-metals which they called 'the *zig-zag* line'. The lesson concluded with the teachers explaining what diatomic elements are and developed their own rules to identify these elements on the periodic table. They called it the '*CONFIB*' rule. This rule states that '*C stands for chlorine, O for oxygen,...and B for bromine....*'

Lesson 2: Interpretation

The lesson focused mainly on the symbolic level of representation. In summary, the lesson underscored the different chemical symbols, the position of the elements on the periodic table (the respective groups), the names of the groups and group numbers and the horizontal periods. Although the lesson included the sub-microscopic aspects, where the link was made between the proton number and the number of protons in the nucleus and the number of electrons, it was disconnected from the wider macroscopic world of real-life phenomena. Indeed, such phenomena (macroscopic level) can only be understood when a learner is introduced to the corpuscular building blocks of Chemistry such as the atom, ions and molecules, which are all embedded in the content derived from the periodic table. In the post-lesson discussion, when the teachers were asked why they had ignored the macroscopic level and its relation to real-world experiences, they said: 'Well, we will have to read up more on that because it's not easy to relate elements to [their] natural occurrences....'

Drawing on Treagust *et al.*'s (2003) typology of verbal explanations, the teachers engaged in several ways with *anthropomorphic* explanations. This happened when they compared the period table with the 'bible', the description of the '*zig-zag*' line for separating metals from non-metals and when they referred to the group eight elements as 'kings' stating that: 'They don't mix with the other elements.' Furthermore, one of the teachers in the group developed a long acronym or rhyme which he described as a key the learners can use to unlock the location of the diatomic elements on the periodic table. These explanations mainly leaned towards analogical understandings as they wanted to show how important the periodic table is when learning Chemistry by comparing it with the bible.

Lesson 3: Acids and Bases

The introduction of the lesson is summarised in the quote below:

> I have just come from the market where I saw lots of fruit and I was wondering how many types of fruit I saw there were acids and bases. I could only afford to buy some oranges, so I want us to test whether oranges are acidic or basic. So, in today's lesson we are going to look at acids and bases. But before we do that, can anyone tell me what is an acid and what is a base?

The class responded:

> An acid tastes sour, Sir, and it has corrosive properties, whereas a base tastes bitter and is not corrosive. For example, a car battery is acidic and it can damage your clothes and skin if it falls on it, whereas the soap that we wash with is a base. It does not damage your skin when you wash with it.

The lesson continued by the teachers explaining the pH scale and how it can be used to differentiate between acids and bases. For example, '…from one to seven is an acid. A base has a pH of above seven. So it falls in between seven and 14…'. This explanation was complemented with different examples that fit within the different ranges of the scale. Here the teachers compared battery acid with vinegar and explained why vinegar can be ingested without causing harm to the human body as opposed to battery acid, which is more corrosive and dangerous and hence it cannot be ingested by the body. From this point onwards the focus shifted towards the different indicators and how they can be used to determine whether a substance can be classified as an acid or a base. After this explanation the class was divided into three groups and provided with an orange and litmus paper, which they had to use to determine whether the juice of an orange can be classified as an acid or a base.

Lesson 3: Interpretation

This lesson focused mainly on the macroscopic and symbolic levels of representation. It also touched on the sub-microscopic properties. For example, the oranges that were collected at the market place, vinegar, battery acid and 'soap' are all associated with the macroscopic levels of representation. The practical activity where the group had to determine the pH of orange juice seemed to shift the focus from the macroscopic level to the symbolic and sub-microscopic levels of representations. When the teachers were asked why they focused mainly on the practical activity they stated that their aim was to do an activity so that the learners can discover the symbolic and sub-micro level of representation on their own.

The model of explanations in Treagust *et al.* (2003) makes it evident that the teachers used a *problem-based approach* to differentiate between acids and bases. Such an approach involves explaining a concept by drawing on a hypothetical problem as a basis to convey the content verbally. Evident in such explanations is the use of calculations, numbers and figures to explain phenomena. This model-based approach was complemented with *relational* explanations in which the teachers used 'battery acid' and 'vinegar' to explain the difference between a strong acid and a weak acid. They also assigned pH values to differentiate between pH strengths, which points towards analogical explanations in which a familiar phenomenon is used to explain an unfamiliar phenomenon. They did not draw on theory to explain these differences, nor did they give any causal scientific explanations for their statements. Instead their definitions of acids and bases were based on the values of the hydrogen ion (pH) and hydroxide ion (pOH) concentrations of the substances. If these explanations are not offered with care, teachers might create confusion or naïve conceptions in Chemistry. One essential component missing from this lesson was the epistemological duty of the teacher to guide the learners to distinguish between the different chemical models of acids and bases by introducing them to their symbolic and sub-microscopic properties. When asked in a post-lesson interview why they defined it in this way, they responded: 'Well, this is how the textbook describes it. Also, learners come to us from lower grades with this knowledge about acids and bases.' The teachers' response is consistent with the view of Van Berkel *et al.* (2009) that textbooks are flawed with misconceptions and inaccurate descriptions. Consequently, teachers must question textbook definitions and descriptions in order to challenge the misconceptions learners have about school science.

Lesson 4: What Are Isotopes?

Owing to spatial constraints only a brief summary of this lesson is provided. The lesson was introduced by drawing an oxygen molecule on flipchart paper as follows:

The teachers then discussed the atomic (proton) number and atomic mass and their relation to the element. For example: 'Can you see the small number and the bigger number in the box.' The continued to explain 'The small number is the atomic number and the big number is the mass number. The number eight means the number of protons found in the nucleus and the difference between the number of protons and the atomic mass number gives us the number of neutrons.' They then draw three different oxygen elements on the flipchart with varying mass numbers and kept the atomic number constant, namely, and explained the difference in the number of neutrons, hence introducing the term isotopes. They continued the lesson by using another element following the same explanation as they did with oxygen.

Lesson 4: Interpretation

The lesson focused mainly on the *symbolic* levels of representation and to a limited extent on the *sub-microscopic* level of representation. The *symbolic level* was used when they introduced the symbols and the sub-microscopic level was used when they provided the subatomic particles such as the mass number that describes the number of protons and neutrons combined and the atomic number indicating the individual number of protons. Furthermore, they used *problem-based* verbal explanations to explain concepts such as neutrons as the difference between the proton number and the mass number. They also used a problem-based approach to explain the term 'isotope' by using calculations and by drawing on the hypothetical representations of oxygen.

Lesson 5: Chemical Bonding

The focus of this lesson was to explain how a chemical bond between sodium and chlorine is formed to produce sodium chloride. They started at the symbolic level by drawing the symbols of the individual elements on the board together with their respective atomic numbers. They then switched to the sub-microscopic level by stating that the number of protons is equal to the number of electrons (seeing that both elements are electrically neutral) and drawing the number of valence electrons around each atom. They then explained the octet rule by stating that 'these elements want to be like the noble gases, they also want to be "kings", therefore they share electrons to form an octet. When chloride has an octet the bond is formed'. They concluded by showing how the bond occurs by sodium transferring its electron to chlorine, resulting in the formation of an ionic bond. They made the link with the macroscopic levels of representation by introducing the molecule as 'that white stuff called salt that you throw in your food to eat, that gives it flavour'.

Lesson 5: Interpretation

In this lesson the focus was on the *symbolic* and *sub-microscopic* levels. This lesson was fraught with misconceptions. For example, they did not mention the words 'electrostatic force of attraction' or 'effective nuclear charge' between the protons and electrons in their presentation. Nor did they give any attention to the binding energy as electrons are bonded to the sodium atom. They created the impression that it is the 'octet rule' that overcomes the binding energy of the electron on sodium resulting in the formation of the ionic bond. They did not mention the fact that lattice energy gives rise to the formation of sodium chloride when millions of chlorine and sodium ions bond through electrostatic interactions.

The explanations that they used were *model-based*, in that they drew a model of a chemical bond and linked it to a *problem-based* approach in which the class had to calculate the number of electrons. They also used *anthropomorphic* explanations when they described the octet rule as 'elements aspiring to be like noble gases' and introducing salt as 'that white stuff that you eat'.

Discussion

Table 4.1 is a summary of how the teachers enacted the three levels of representations and the types of explanations they used in each lesson.

Table 4.1: Summary of lessons and the enactment of triplet Chemistry and types of explanations

Les- son	Level of rep- resentation			Type of explanation					
	Macro	Symbolic	Sub- micro	Ana- logical	Anthro- pomor- phic	Rela- tional	Problem- based	Model- based
1		Y	Y			(Y)		Y
2		Y	(Y)	Y	Y	Y		
3	Y	Y	(Y)		(Y)	Y	Y	
4		Y	Y				Y	
5	Y	Y	Y		(Y)		Y	Y

Y= identified by you; (Y) = identified somewhat by you.
Source: Author.

By linking the five lessons with the way the teachers enacted the three levels of representation and explanations, the findings reveal that even within a professional development programme emphasising the interconnectedness of the three levels of representation, none of these groups were successful in such integration in their lessons. The findings further reveal that the emphasis is more on the symbolic and sub-micro level and little connection is made with the macroscopic level of representation. This contrasts earlier findings by Treagust *et al.* (2003) which seem to privilege enactment at the macroscopic level, while the teachers could not shift between all three levels of representation.

Each lesson, irrespective of which level of representation, was emphasised, revealing misinterpretations and misconceptions of various key terms as well as of the models they used to explain phenomena. The findings from this study and similar other studies (Davidowitz, Chittleborough, & Murray, 2010; Taber, 2013; Talanquer, 2011) are a cause of concern about the way teachers teach basic chemistry to their learners. For example, in Lesson 1 the model of the atom was presented as electrons orbiting the positively charged nucleus. This model, Talanquer (2011) explains, was chosen on the basis of the teachers' visualisation of the atom. Although this model (Bohr's model) is not incorrect, it lacks rigorous scientificity since in science there are always competing models. For example, in the hydrogen model (also referred to as the single electron model) of the atom the electron does not move around the nucleus in discrete energy lanes, but moves randomly around the nucleus and is kept intact by the effective nuclear charge of the atom.

Possible reasons for the focus on the symbolic level in this study could be the assumed progression by learners from earlier grades and the fact that most teachers focus more on preparing their learners for the National Senior Certificate examinations. Another reason could be the fact that the focus on most chemistry topics in textbooks is on the symbolic level and sub-microscopic levels with little consideration given to the macroscopic properties and its applications in real life. Treagust *et al.* (2003) point out that effective learning of chemistry requires teachers to shift between all three levels in chemical explanations. These three levels, Talanquer (2011) notes, are a complex knowledge space, which makes chemistry a complex subject. He points out that when teachers teach any topic in chemistry, it is pedagogically advantageous to begin with the learners' understanding of a topic by drawing from their personal experiences of the real world (macro level). This can be followed by introducing them to various models by highlighting the scientific concepts used to build the models (symbolic level). The findings also revealed that the teachers relied more on *anthropomorphic*, *relational* and *problem-based* explanations and to a limited extent on *model-based* and *analogical* explanations to deliver the content. These types of explanations have the potential to create confusion and misconceptions about scientific concepts if learners cannot relate them to real-life examples to explain phenomena. Because of the complexity of triplet chemistry and to avoid learner confusion and misconceptions with these types of explanations, it is important that teachers should not overload learners with content during a lesson (Dumon & Mzonghi-Khadhraoui, 2014; Taber, 2013). By restricting the amount of information conveyed to the learners, they can interrogate and integrate the core content in a more robust manner. This will assist learners to make the link between existing information and the new information provided at the respective levels by the teacher. When learners grasp the new content at the macro and symbolic levels, teachers can then model the way chemists operate between the three domains of triplet chemistry by using the language of the sub-microscopic level. During these explanations teachers must use precise language and offer learners sufficient scaffolding to assist them in gradually learning to operate within and across the various levels of representation to avoid creating epistemological obstacles.

Recommendations and Conclusion

The findings revealed that the professional development programme did not succeed in what it set out to achieve, that is, the integration of the three levels of representation in chemistry. This implies that one could question the usefulness of calling on the teachers' own practices tempting them to present what they refer to as an 'exemplary lesson'. It is therefore recommended that teachers can design a new lesson 'together' by following the pedagogical steps outlined in the findings to assist them in the planning of a lesson, particularly in addressing the shift between all three levels of representation. They can then try it out in their classrooms and report back through mobile phone footage.

The findings also have implications for curriculum development in South Africa. It is therefore recommended that the detailed CAPS curriculum includes frequent cross-references and opportunities for linking the three levels of representation explicitly. This will assist teachers to overcome the challenge that the teaching of Chemistry faces in South Africa by evolving suitable didactic practices to preserve the unity of chemistry at all three levels of representation.

References

Andersson, B. (1986). Pupils' explanations of some aspects of chemical reactions. *Science Education, 70,* 549–563.

Basson, I., & Kriek, J. (2012). Are grades 10–12 Physical Science teachers equipped to teach Physics? *Perspectives in Education, 30*(3), 110–122.

Bodner, G. M. (1992). Refocusing the general chemistry curriculum. *Journal of Chemical Education, 69,* 186–190.

Bucat, B., & Mocerino, M. (2009). Learning at the sub-micro level: Structural representations. In J. Gilbert & D. Treagust (Eds.), *Multiple representations in chemical education: Models and modelling in science education* (pp. 11–29). London: Springer.

Cheng, M., & Gilbert, J. (2009). Towards a better utilization of diagrams research into the use of representative levels in chemical education. In J. Gilbert & D. Treagust (Eds.), *Multiple representations in chemical education: Models and modelling in science education* (pp. 55–73). London: Springer.

Davidowitz, B., Chittleborough, G., & Murray, E. (2010). Student-generated submicro diagrams: A useful tool for teaching and learning chemical equations and stoichiometry. *Chemistry Education Research and Practice, 11*(3), 154–164.

Department of Basic Education. (2011). *Curriculum and assessment policy statement: Further education and training band.* Pretoria: Author.

Dumon, A., & Mzonghi-Khadhraoui, S. (2014). Teaching chemical change modelling to Tunesian students: An 'expanded chemistry triplet' for analysis in teachers' discourse. *Chemistry Education Research and Practice, 15*(1), 70–80.

Gilbert, J., & Treagust, D. (2009). Introduction: Macro, sub-micro and symbolic representations and the relationship between them: Key models in chemical education. In J. Gilbert & D. Treagust (Eds.), *Multiple representations in chemical education: Models and modelling in science education* (pp. 11–19). London: Springer.

Harrison, A. G., & Treagust, D. F. (1996). Secondary students' mental models of atoms and molecules: Implications for teaching chemistry. *Science Education, 80*(5), 509–534.

Harrison, A. G. and Treagust, D. F. (1998). Modelling in science lessons: are there better ways to learn with models? *School Science and Mathematics, 98,* 420–429.

Johnstone, A. H. (1993). The development of chemistry teaching: A changing response to a changing demand. *Journal of Chemical Education, 70*(9), 701–705.

Kozma, R. B., & Russell, J. (1997). Multimedia and understanding: Expert and novice responses to different representations of chemical phenomena. *Journal of Research in Science Teaching, 34,* 949–968.

Locke, J. (2009). *Of the abuse of words.* London: Penguin Books.

Onwu, G., & Randall, E. (2006). Some aspects of students' understanding of a representational model of the particulate nature of matter in chemistry in three different countries. *Chemistry Education Research and Practice, 7,* 226–239.

Ramnarain, U., & Joseph, A. (2012). Learning difficulties experienced by grade 12 South African students in chemical representation of phenomena. *Chemistry Education and Practice, 13*, 462–470.

Rogan, J., & Grayson, D. J. (2003). Towards a theory of curriculum implementation with particular reference to science education in developing countries. *International Journal of Science Education, 25*(10), 1171–1204.

Rollnick, M., Bennett, J., Rhemtula, M., Dharsey, N., & Ndlovu, T. (2008). The place of subject matter knowledge in pedagogical content knowledge: A case study of South African teachers teaching the amount of substance and chemical equilibrium. *International Journal of Science Education, 30*(10), 1365–1387.

Stott, A. (2013). South African physical science teachers' understanding of force and the relationship to teacher qualifications, experience and their school's quintile. *African Journal of Research in Mathematics, Science and Technology Education, 17*(1), 173–183.

Taber, K. (2013). Revisiting the chemistry triplet: Drawing upon the nature of chemical knowledge and the psychology of learning to inform chemistry education. *Chemistry Education Research and Practice, 14*(2), 156–168.

Talanquer, V. (2011). Macro, submicro and symbolic: The many faces of the chemistry 'triplet'. *International Journal of Science Education, 33*(2), 179–195.

Treagust, D., Chittleborough, G., & Mamiala, T. (2003). The role of sub-microscopic and symbolic representations in chemical explanations. *International Journal of Science Education, 25*(11), 1353–1368.

Umalusi. (2011). *Annual report: Strength, growth and stability*. Department of Basic Education. Pretoria: Author.

Van Berkel, B., Pilot, A., & Bulte, A. (2009). Micro–Macro thinking in chemical education: Why and how to escape. In J. Gilbert & D. Treagust (Eds.), *Multiple representations in chemical education: Models and modelling in science education* (pp. 31–55). London: Springer.

CHAPTER FIVE

Argumentation as a Teaching Method

Introduction

In post-apartheid South Africa, where there has been a concerted effort towards increasing the number of scientists, engineers and technologists, it is undoubtedly teachers that have a great role to play towards achieving this goal. In the present South African basic education curriculum (Department of Basic Education [DBE], 2011), the main focus is on understanding the nature of science, technology and other phenomena in everyday life. The Curriculum and Assessment Policy Statements—CAPS (DBE, 2011)—propose the development of 'critical thinking' as one of the goals. This is a goal that can be highly beneficial to teachers and learners when they identify, evaluate and craft scientific arguments. Among the skills which learners are meant to acquire after studying physical sciences at high school is the ability to read about and understand scientific phenomena, to develop critical thinking skills and to develop scientific reasoning processes, problem-solving skills and strategic abilities. Since it is an aim of the CAPS that teachers and learners be conversant with how to evaluate scientific arguments, it is important that we rethink what is being taught and how it is being taught.

Implementation of the CAPS has been reported to have had a marginal impact on learners' performance in that the system of basic education is said to be on the right path as it moves towards higher standards and improved quality

(DBE, 2011). Despite this sense of relief after a deluge of criticism against the Department of Education (DE), not much change has occurred relative to the significant decline in Grade 12 physical science pass rates over the last four years. There has been a steady increase in the number of candidates who registered for the Physical Sciences examination in the last four years (2013–2016)—from 184,383 to 192,618. Furthermore, in this period the number of physical science candidates who pass at the 30-percent level has decreased from 124,206 to 119,427 (Department of Education [DoE], 2016, p. 50). The national pass rate, however, has declined from 67.4 percent to 62 percent. This decline in the number of Grade 12 candidates passing physical sciences in the last four years has raised many concerns amongst stakeholders throughout the country. Several of these concerns revolved around what is being taught and how it is being taught. Teacher talk and learner passivity still hold sway in the science classroom. Yet the prominence given to dialogue, argumentation, discussion and group activities in the South African curriculum stands out in sharp contrast to the implementation of a traditional instructional approach and the rote learning associated with the old curriculum (Ogunniyi, 2007a). The contention is therefore that the attainment of the new curriculum objectives is fundamental to improving the situation, but that this requires strategic innovation and training to achieve the set goals.

As a way of addressing these troubling issues in the teaching and learning of sciences in South Africa, a very particular didactic approach has to be adopted. There is a need for the consistent application of sound teaching methods promoted by the new curriculum document, CAPS. This process may be facilitated by the learner-centred approach in a cooperative learning environment mandated by the new curriculum.

Various authors have shown that interactive classroom argumentation and dialogue have tended to encourage teachers and learners to externalise their views on any subject matter (Erduran, 2007; Jimenez-Aleixandre& Erduran, 2008; Kuhn, 2010). For this to be realised, the teaching and learning of science should move away from traditional patterns. Lemke (1990) argues that traditional patterns of teaching science place teachers in a position of power in which they control the teaching and learning process. In the same vein, Lehrer and Schauble (2006) assert that traditional science discourse patterns are not appropriate as the sole discourse pattern in inquiry-oriented classrooms, because they are based on teacher-driven instruction and known answers to questions. What seems needed therefore is an emphasis on effective teaching methods to integrate epistemological issues in science teaching and learning. Effective teaching methods encourage an atmosphere where ideas are raised and then confirmed or contradicted by evidence and by the arguments of others (Lawson, 2004). This would lead to a better understanding of specific parts of science and a deeper awareness of what science is as a discipline. This is most important, especially today when there is much concern about the level of science that learners are learning and about their decreasing interest in the field at a time when the need is rising for science and technological skills as well as a broader general education.

We live in extraordinary times where science is revealing truths about the universe that are both staggering and breathtaking. The most up-to-date findings of physics suggest there is no space between separate objects, or indeed that there are no objects as we normally think of them, and that the whole notion of 'space' or 'separate objects' has no foundation in reality. And that everything consists of vibrating strings of energy. Yet, as incredible and fantastic as all this sounds, perhaps the overwhelmingly important question to ask is: How do we as science educators talk about events or relational concepts to our learners, if we must give up all conceptions of time and space, and never say that one thing causes another to happen?

Below I provide a detailed narration of the difficulties I experienced in the teaching of selected topics from the Physical Science syllabus and how argumentation as a teaching method assisted me in addressing these challenges.

My Personal Lived Experiences as a Physical Science Teacher

Of immediate relevance to the above discussion is the account of my personal experience in working with high school and university learners in South Africa. My first post, where I taught for most of my career as a Physical Science teacher, was in a township school in Cape Town. Like most 'township schools' in the city of Cape Town the entire learner population is black, their mother tongue is IsiXhosa—although the medium of instruction is English—and the teaching staff is entirely black. Furthermore, given the fact that the school is a 'no fees school', it is poorly resourced and most of the learners come from single-parent families and have to deal with extreme poverty. Most of them live in shacks, which are makeshift shelters with sometimes no electricity. These learners have strong cultural values and are raised to honour a culture-bound framework of thinking. As an African foreign national growing up in Nigeria, my classroom experience provided me with a unique opportunity to encounter the real difficulties in teaching certain science concepts to my learners. My learners often lamented that Western science, with its universal remit, contradicts their own belief system and modes of coming to know something. For this reason they would never hesitate to say 'Sir, science is not for us.' Why? I asked. 'Sir, what is scientifically valid may not always be considered culturally valid; we have our own way of relating to science and the world. It involves personal transformation.'

This experience nearly halted my science teaching career during my first year at the secondary school. Even though I had graduated from the university with flying colours in science and mathematics education, my learners' worldview seemed to represent a steep mountain to climb. I became despondent. I thought all my hopes for my teaching career were lost. I was stuck with the traditional method of teaching science, the so-called chalk-and-talk approach. Indeed, I was struggling to find a better way to help my learners embrace what Ausubel (1968) calls meaningful learning. My learners being English second-language speakers also faced the challenge of understanding the language of science and were not ready to accept that the only truth about the world is that which is advanced by Western science. In addition to all these challenges, my school pressured me to perform a miracle over the poor pass rate in Physical Science at the school. Countless times the education district officials visited my school to inform the school principal about their intent to withdraw the school's status as a 'Dinaledi school' (i.e. science and mathematics focused school). This had been an ongoing concern before my arrival at the school. Central to this concern is the condition the school placed on my first contract to help them turn their final-year Grade 12 Physical Science results around so that they could continue to enjoy the benefits of their prestigious status.

Indeed, my learners were struggling, and so was I, in my attempts to help them to develop a good understanding in order to encourage a perceptual shift about the nature of science. Some examples of the difficulties I encountered in my teachings include learners' conceptual ecology that led to difficulties over scientific claims about lighting and thunder, which form part of the concept of electrostatics and so on. A person's conceptual ecology, according to Hewson (1996), is what he/she uses to determine whether certain conditions for truth/validity are met, whether a new conception is intelligible or makes sense, whether it is plausible or can be believed to be true and fruitful or useful. However, if the new concept conflicts with existing concepts, then the new conception cannot become plausible or fruitful until the learner becomes *dissatisfied* with the old concepts.

Another difficulty I noticed among my learners was their resistance to conceptual change and their tendency to construct what they considered to be common-sense-based knowledge. This challenge compelled me to explore various instructional strategies until I found myself very much preoccupied with argumentation as a teaching method. I thought to myself that I should be able to make a change in the lives of my learners, including improving the Physical Science results at the school. So I decided to put argumentation theories to the test. I must admit there was a feeling of both excitement and trepidation with this decision. What if it doesn't work? This would disprove everything I believe in, my self-efficacy and expectations. If it doesn't work, then better to move on and not waste any more of my time and that of my learners; but I dismissed my doubts and resolved to take action. I endeavoured to plan all my lessons and related activities using an argumentation-based scheme. Every activity I planned involved argumentation as a scaffold for my learners. I moved away from the traditional chalk-and-talk pattern to something new.

As a science and mathematics educator, I first discovered the argumentation framework more than a decade ago. I am certain that this early introduction influenced both my teaching career and my way of looking at the didactic implications of argumentation. In recent years argumentation instruction has brought a fresh perspective into the teaching-learning process, especially in Africa, where the traditional chalk-and-talk approach dominated classroom discourse for decades. Since I discovered argumentation as an instructional tool in my teaching of science to learners, it has always offered them a great deal of interest and insight towards achieving a better understanding of scientific phenomena. My learners who used to say 'science is not for us' started to find their learning of science meaningful. The Physical Science pass rate at my school went from being very poor to average, and two years later the pass rate drew the attention of district education officials. I was honoured at the school and invited by the district officials to do a presentation for nearly 120 guests (teachers, curriculum advisors, etc.) on how I produced such results and turned the situation at the school around. For the first time science learners in my school achieved outstanding performances of more than 70 percent. A few examples of my teaching of science at this school through argumentation in a science classroom will be discussed later in this chapter. But first I want to differentiate between *argument* and *argumentation*.

Differentiating Between Argument and Argumentation

The meaning of 'argument' in educational literature may be understood in two ways, generally speaking. One definition, according to the Britannica (2008), is that argument entails advancing a reason for or against a proposition or course of action. This kind of argument is common in science lessons in which a teacher comes to a class with a scientific explanation and helps learners to see it as reasonable. The second interpretation of argument is that it is 'dialogical' engagement, which entails the examination of different perspectives and the purpose is to reach agreement on acceptable claims or courses of action (Driver, Newton, & Osborne, 2000).

Argumentation, on the other hand, according to Kuhn and Udell (2003), is the process of making a claim and using evidence to justify that claim. To make more sense of these two constructs Toulmin (2003) explains that argument refers to the substance of claims, data, warrants and backing that contributes to the core content of the argument. Argumentation, however, refers to the process of assembling these components. The main difficulty has been in attempting to clarify what counts as a claim (C), data (D), warrant (W) and backing (B).

In his thought-provoking and insightful book entitled *The Uses of Arguments* Toulmin diligently explains that when a *claim* (C) is being made, it can be challenged by a questioner who asks 'What reasons have you got to go on?' In this case the person who made the claim can then appeal to the relevant facts at his/her disposal, which Toulmin calls *data* (D).Even when we have the correct facts at our disposal, he contends, it may turn out to be necessary that we establish them in a preliminary argument. Whether they are accepted by the challenger or not does not necessarily end the argument.

Consequently, an alternative scenario is that the challenger may ask about the bearing of our data on our claim. This gives rise to the question: 'How do you get there?' In this case our response will take the form: 'Data such as D, which entitles us to draw conclusions or to make claims such as C.' This form of proposition is what Toulmin calls a *warrant* (W) (p. 98). To him, warrants (W) confer different degrees of force on the conclusions they justify, which may be signalled by qualifying our conclusions with a *qualifier* (Q) with cues such as 'necessarily', 'probably' or 'presumably'.

What Toulmin means by conferring different degrees of force on the conclusions is that 'the force of the term "cannot" include, for instance, the implied general injunction that something or other has to be ruled out in this or that way and for such a reason"(p. 28). Hence a counter condition may arise during the arguments may arise: Toulmin calls this a *rebuttal* (R). Conditions of rebuttal must be mentioned, in a best-case scenario 'indicating circumstances in which the authority of the warrant is set aside' (p. 101).

Furthermore, the challenger may again question the general acceptability of our warrant (W) by asking: 'Why do you think that?' Thus, the person who made the claim will have to provide an answer that substantiates his/her thought. Such an answer is what Toulmin calls the *backing* (B). Toulmin further emphasises various forms of backing in different fields. For example, he explains that warrants can be defended by appeal to a system of taxonomic classification, to a statute, to statistics from a census and so on. It is this difference in backing that constitutes what he calls the field-dependence of standard arguments. To clarify these standards, Toulmin explains that they comprise 'the grounds and reasons, by reference to which we decide in any context that the use of a particular modal term is appropriate'. He writes:

> Two arguments will be said to belong to the same field when the data and the conclusions in each of the two arguments are, respectively, of the same logical type: they will be said to come from different fields when the backing and/or the conclusions in each of the two arguments are not of the same logical type (Toulmin, 2003, p. 14).

To this end, all micro-arguments depend on the combination of data (D) and backing (B). If it is desirable that our backing is to be checked, then it will involve checking our claim too. Toulmin calls such arguments 'analytic arguments'. It must be pointed out that arguments of this kind are rare and most arguments are not of this sort. Arguments that fall within the ambit of formal criteria Toulmin calls 'substantial arguments'. Any claim made must be backed up, hence there is a wide variety of types of claims to be examined. Anyone's claim has the right to be taken seriously and examined on its own merits. Of this claim we are entitled to say that some possibility has to be ruled out only if we can produce grounds or reasons to justify this claim, and under the term 'criteria' can be 'subsumed the many sorts of things we have then to produce' (Toulmin, 2003, pp. 28–29).

In summary Toulmin claims that any examination must contain the following stages:

- We state the problem;
- We acknowledge that different solutions are possible;
- We consider those that we deem possible (this possibility is not the 'possibility' of logic; it simply means that we think they are worthy of consideration);
- We come to a decision.

Despite the frequent use of Toulmin's argumentation pattern (TAP) by science educators, criticisms directed at it have not ceased. For example, there have been questions around the issues of what the quality of argument is and ways by which to determine its strength. Toulmin (2003), in agreement with various scholars, avers that the quality of arguments should not be judged on the basis of individual components, but rather on their overall structure (Berland & Reiser, 2008; Chinn, O'Donnell, & Jinks, 2000; Clark, Sampson, Weinberger, & Erkens, 2007).

Exceptions to this universalistic stance in science education research are the works of researchers who are conversant with the strengths and limitations of the model and hence have made painstaking efforts to adapt the model where feasible. In their work with teachers and learners, some of these researchers referred to earlier have found that there is no common pattern in the way teachers use the model or even the same form of arguments in their classrooms. In other words, the use of argument appears to be teacher dependent (Erduran, Simon, & Osborne, 2004; Jimenez-Aleixandre, Rodrigues, & Duschl, 2000; Kelly & Takao, 2002; Ogunniyi, 2007a, 2007b; Simon, Erduran, & Osborne, 2006; Zohar & Nemet, 2002). It is the researcher's belief that what lies behind this relates largely to the fact that the application of TAP to the analysis of classroom-based discourses has yielded difficulties (if not more questions than answers). To explain these shortcomings I now turn my focus to contemporary scholars who have managed to address these shortcomings.

Addressing the Shortcomings of TAP

Erduran *et al.* (2004) alluded to the difficulty of distinguishing various components of Toulmin's argument, particularly data, warrants and backings, which were had already been revealed by other scholars (Kelly, Druker, & Chen, 1998). In their attempt to provide a methodological pathway while using TAP as strategy for teachers to instil scientific reasoning skills in their learners, they simplified the elements of TAP by designating data, warrants, backings and qualifiers as grounds. This arrangement immediately removes the contention that has been raging around the elements of TAP (Ogunniyi, 2007a).

In the South African context a case study conducted by Ogunniyi (2007a) has identified further TAP shortcomings. The author states:

> ...another problem with the TAP is whether or not we can accept a set of data forming the basis of a claim at face value without considering the underlying assumptions or theoretical constructs that give such data specified meanings. (p. 966)

I think what has been gained from these authors in the use of TAP is that several questions remain concerning its methodological constructs, hence the overlapping elements such as data and warrant, as pointed out by Van Eemeren, Grootendorst, and Kruiger (1987). Toulmin's argumentation pattern (TAP) offers a useful tool for arguing in informal settings such as in socio-scientific and scientific contexts, where informal logic is applicable. However, the way TAP has been applied in many studies has been criticised in terms of the inconsistent means by which the validity of an argument is established (Van Eemeren *et al.*, 1987).

Ogunniyi (2007a) outlined different types and levels of arguments, serving a variety of functions, concluding that it is not feasible to use a single model to represent all forms of arguments. Based on these insights, various authors have put forward convincing arguments to the effect that teaching science through collaborative or dialogic argumentation, in which both logical and non-logical issues that often arise in multicultural classroom discourses are considered, would lead to meaningful learning (Asterhan & Schwarz,2007;Diwu & Ogunniyi, 2012; Ogunniyi, 2004, 2007a,2007b). I am convinced that, in science classrooms, it is important not only for learners to be able to make sense of data to construct claims, but also to be able to consider alternative claims and to critique the claims and justifications provided by their fellow learners in the context of dialogic interactions. As stated earlier, argument and the argumentation practice are seen as a core activity of scientists and have a central role in science education. To enhance the public understanding of science and to improve scientific literacy, argumentation must be given a high priority in education in science and within science itself (Driver *et al.*, 1998). This raises an important and poignant question: Should we continue to teach science concepts without exposing our learners to a dialogical discourse about the existing alternative ideas about certain science phenomena?

From the history of science perspective, science has tended to progress more by argumentation, dialogues and revolutionary ideas than by consensus (Kuhn, 1970; Popper, 1968). Therefore, it is important that learners engage in dialogical argumentation and develop their argumentation and reasoning skills to understand science, themselves, others and the world at large better. The importance of the role of dialogical argumentation in learning has been obvious in research works in the field of science education for a long time (Wandersee, 1985).Research has shown that the teaching of argumentation through the use of appropriate activities and pedagogical strategies can be a means of promoting epistemic, cognitive and social goals as well as enhancing learners' conceptual understanding of science (Erduran *et al.*, 2004). It is not difficult to understand why the authors held such a view. It can be said that the adoption of any new approach that promotes the use of argument would require a shift in the nature of the discourse in science teaching and so is the central aim of this chapter. In this regard, Dialogical Argumentation Instructional Model (DAIM) espoused by Ogunniyi (2009) is a useful instructional tool for implementing dialogical argumentation in classroom. The immediate advantages have been the opportunity created through its various stages to undertake dialogical argumentation that consider both logical and non-logical issues that often arise in multicultural classroom discourses.

Pedagogical Scheme for Implementing DAIM

DAIM involved five stages of learning. The goal of learning through DAIM is to engage learners in legitimate peripheral participation in communities of practice (Lave & Wenger, 1991). Through a community of practice, learners interpret, analyse, reflect and form meaning about a given task. I will now provide an explanation to show the interrelatedness of each stage of DAIM, as depicted in Figure 5.1.

Figure 5.1: The five stages of implementing DAIM.

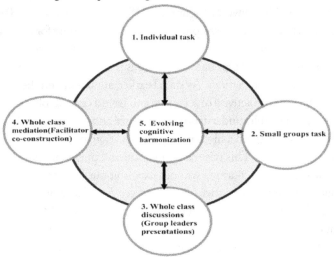

Source: *Paper presented at the second national workshop on science and indigenous knowledge system at the University of the Western Cape,* Ogunniyi (2009).

Stage 1: When an individual is confronted with a task (or given problem), a sort of 'internal argument' or 'conversation' ensues within him/her (at the microneuro-psychical level) where consciousness in an individual is assumed to be most active (Ogunniyi, 1997).

Stage 2: This is the stage where individual thinking or self-conversation (intra-argumentation) is brought into a social setting involving a small group of people arguing and discussing a given task (inter-argumentation).

Stage 3: At this stage group leaders or representatives present the decision reached by their respective group to the whole group for further debates and discussion.

Stage 4: At the fourth stage, the facilitator (teacher) mediates the discussions. This is trans-argumentation.

Stage 5: At the fifth stage, after stages 1 to 4 have been successful, an evolving cognitive activity is already in progress in the mind of the individual as arguments and counter-arguments are presented.

As already found elsewhere during the joint construction of a solution enacted through arguments, individual group members can request explanations and justifications from one another in well-functioning groups. Learners share their intellectual strategies as together they solve a given problem (Heller, Keith, & Anderson, 1992). Results from the authors further suggested that better solutions to problem emerged through collaboration than were achieved by individuals working alone. As has been noted by others (Ogunniyi, 2007a, 2007b), successful implementation of DAIM is capable of achieving these objectives. One reason for this enthusiasm is that many studies have found that argumentation instruction allows learners to participate more freely in classroom discourse, clear their doubts and even change their views than would otherwise have been the case with traditional teacher-dominated instruction (see, e.g. Erduran *et al.*, 2004; Ogunniyi, 2007a,2007b; Osborne & Patterson, 2011).It is therefore noteworthy that the main idea behind the successful use of DAIM is based on how effectively it is implemented. In view of its effectiveness, DAIM could be used to revive learners' declining interest in learning science as it provides the setting for the social interaction needed by learners to share tacit understandings and to create shared knowledge from the experiences among participants in a learning opportunity (Wenger, 1998).

Besides the implementation of DAIM, the question of what can or cannot be debated in classrooms remains open and is not easily answered nowadays. In this regard, efforts should be made to ensure all arguments are non-intimidating, a necessary attunement for multicultural South African classrooms. Based on my experience in working with science and mathematics learners at both basic and higher education levels, very often learners' utterances in class are fragmented and overlapping. Science classrooms are sometimes filled with messiness, which makes it difficult to capture group dynamics and chains of reasoning as groups construct and evaluate arguments pertaining to a given task. About these considerations people differ in the areas in which they cultivate their arguments and the levels to which they develop them even within their given pursuits. For example, an African science learner who uses non-logical elements to foster his argument may see his insight disregarded in space (logic context) and that diminishes (or silences) his worldview. Thus, he might feel isolated and be discouraged from carrying on with the dialogic argument.

Supplementing this, Ogunniyi's (2004, 2007a) contiguity argumentation theory (CAT)caters for both logical and non-logical aspects of argument worth using to present fresh perspectives on how African science learners use the nature of argument to construct their knowledge in science discourse as opposed to having an argumentation framework that caters only for monologues.

Contiguity Argumentation Theory (CAT)

Essentially, Ogunniyi's (2004, 2007a, 2007b) CAT recognises five categories into which ideas can move within a person's mind when discussing issues of different thought systems. Although CAT to some extent draws on the Aristotelian notion of contiguity, it regards such elemental ideas not as 'concrete referents' but as dynamic organising conditionals or 'frames of reference' that galvanise the process of association or learning in general, depending on the context in question (Ogunniyi, 2007a, p. 162). Also, it is evident that CAT is universal in its application and has educational implications for teachers at all levels of education. Likewise, CAT draws on the African worldview theory of *ubuntu*, which stresses the interdependence or interrelatedness of ideas. In other words, ideas find full expression and authentication in the collective ideas of others or the society at large.

Furthermore, CAT is a handy analytical tool for interrogating classroom discourses. Before CAT became well known, most scholars usually considered Western science and indigenous knowledge systems (IKS) as incompatible. CAT has brought to scholars' awareness the legitimacy of indigenous knowledge as a valid way of interpreting experience. The immediate advantages of CAT have been to provide valuable theoretical underpinnings and successful instructional strategies from research for learners coming from indigenous backgrounds. CAT identifies five cognitive states: dominant, suppressed, assimilated, emergent and equipollent. During each of these stages there is a unique level of analysis, internal organisation and understanding of the cognitive shift that occurs in the mind of the learner. These five adaptive cognitive states could occur within a learner confronted with counter-intuitive school science. The five adaptive cognitive states are summarised in Table 5.1.

Table 5.1: The five stages of CAT as proposed by Ogunniyi

CAT Stages	Adaptive co-existence that occurs within a learner confronted with two distinct forms of thought/constructs (*logical and non-logical features*)
(1) Dominant	A thought system or an explanation is seen to be more convincing or more appropriate than any other thought system or explanation at that moment and for that context, or a powerful idea explains facts and events more effectively and convincingly than another idea, or it resonates with the acceptable social norm that affords an individual a sense of identity.
(2) Suppressed	A thought system is seen as less convincing than another. The less convincing, subordinate or culturally accepted thought system becomes suppressed. An idea becomes suppressed in the face of more valid, appropriate, adequate and convincing evidence.

(3) Assimilated The dominant thought system is adopted by people who initially held the suppressed thought system as the standard. The initially held thought system is supplanted or subsumed by the dominant worldview. This means that a less powerful, less convincing idea is assimilated (taken in, swallowed by) a more powerful or more persuasive idea.

(4) Emergent No previous idea, opinion or position on an issue really exists in the learner. For instance, an idea may emerge as the individual is exposed to new teaching, as is normally the case with science concepts learnt at school. The learner then adds the new concepts to his repertoire of knowledge.

(5) Equipollent Two competing thought systems are seen as equally powerful, adaptable, active, effective or significant in making sense of the observed phenomena. The two rival thought systems coexist and exert equal cognitive force on a person's beliefs. For example, a person could find both theories of creation and evolution as attractive explanations for the origin of the universe and humankind.

Adapted from: *International Journal of Science Education*, *29*(8), Ogunniyi (2007a).

Each of these cognitive dimensions is in a dynamic state of flux and can change from one form to another depending on the context in question (Ogunniyi, 2007a). To buttress the role that CAT plays during making claims and counter-claims, Ogunniyi explained, CAT holds that claims and counter-claims on any subject matter within (or across) fields can only be justified if neither thought system is dominant. Hence there must be valid grounds for juxtaposing the two distinctive worldviews within a given dialogical space. Consequently, the role of such a dialogical space is to facilitate the process of re-articulation, appropriation, or what Ogunniyi calls negotiation of meanings of the different worldviews (e.g. science and IKS). Learners must therefore be able to negotiate the meanings across the two distinct thought systems in order to integrate them. This is because CAT assumes that ideas that come together will interact, overlap or conflict with each other. In other words, when ideas clash, an internal dialogue occurs to find some meaningful form of coexistence. One way of integrating such systems of thought is by finding or adapting to a larger synergistic milieu of conceptualisation (Ogunniyi, 1997). To put the above points into perspective, I have provided an exemplar of what teaching dialogical argumentation to non-Western science learners (e.g. African learners) who hold a different worldview might look like. I have also briefly indicated the roles of learners and teachers in a science classroom that draws on dialogical argumentation.

Instructional Strategies for Promoting Dialogical Argumentation

Proper planning is needed for promoting DAIM as an instructional strategy. I suggest the teacher or facilitator should first determine the intended learning outcomes. Differentiate between the intended outcomes and learners' conceptions (including nature of conflicts). Design a lesson with specific tasks (e.g. an experiment or a worksheet) that lead towards the attainment of the intended outcomes. I have provided few steps to follow in teaching electrostatics to learners who hold different worldviews from orthodox science.

Teaching the concepts of electrostatics can be challenging to many African science learners, who believe that lighting comes from other sources than those identified by Western science. In South Africa lightning is called *kuhambele umhlekazi*, which is believed to be an important messenger from the ancestors—an honoured, respected visitor from high levels. In other African countries such as Nigeria, which has deeply rooted cultural perspectives, lighting is called *amuma miri*, which is believed to be the spirit of the rain. In South Africa those affected by the lightning know that they must have offended the ancestors and must repent and restore their relationship with them. Likewise, many African science learners in South Africa believe that thunder is an elderly mother sheep and her son is lightning. The son is short-tempered and quickly destroys houses and property when angry. His mother would then raise her voice to control him, but he is always too fast for his elderly mother.

In Nigeria thunder is called *Amadioha*, which is interpreted as god of thunder in the likeness of a big male sheep with two powerful horns in front. This god of thunder strikes people who offend him, but can only do so when invited by anyone who pay homage to him. Others believe it stops at nothing when provoked. It can kill and destroy anything that stands in his way, including human, plants, crops, trees and houses. On the other hand, Western science informs us that during a thunderstorm the clouds and the ground acquire negative and positive charges. When the negative charges in the cloud become too great or huge for the cloud, there is a discharge of energy from the cloud to the ground, which we call lightning. As the charge moves to the ground it displaces air molecules violently resulting in the sound we call thunder. It is at this point that a tantalising paradox emerges, which I narrated earlier when I faced teaching science concepts to my learners. Based on my insights gained from that experience, I present an example of dialogical argumentation which I think can reveal ways in which non-Western science learners holding different worldviews can begin to engage in dialogues with the respect and courtesy that is the hallmark of humanity.

In a DAIM teaching situation, the teacher's role is that of a facilitator and an organiser of instructional sequences with the sole aim of guiding learners towards the intended outcomes (Ogunniyi, 2009). It is unhelpful if the teacher creates a conceptual conflict or a dialogical-argumentation situation and then leaves learners entirely to their own devices. Learners need the necessary support through modelling and scaffolding to reach the intended outcomes.

Step 1: Design a specific task enriched with conflict

The design of the learning tasks is threefold. The learning tasks should have the characteristics of DAIM and aim at: (1) determining learners' conceptions; (2) destabilising such conceptions e.g. by presenting or confronting them with conflicting evidence or conceptions; (3) gradual reconstruction on the part of learners in the direction of the intended learning outcome. Encourage learners to discuss their opinions with other members of their groups (Ogunniyi, 2004). An example of a teaching-learning context to enhance learners' conceptual understanding is illustrated below.

Electrostatics (The study of charge behaviour—a controversial topic for African science learners).

Lightning is an electrical discharge in the atmosphere. The very large and sudden flow of the charge that occurs in lightning has enough energy to kill people or do serious damage to buildings or infrastructures. In many traditional beliefs, lightning can come from other sources.

Scenario: **A man is followed and killed by lightning after a quarrel with another man.**

This incident is said to have taken place in a certain African village. Two men, Okonkwo and Achebe (pseudonyms), quarrelled at a beer parlour. Okonkwo threatens Achebe but does not specify the nature of the punishment. A few days later Achebe's homestead is struck by lightning, burning a hut. Achebe is not at the homestead. He is at a beer parlour not far from his home. Although nobody is killed by the lightning, those present are terrified by the intensity of the damage caused by the lightning. Achebe sees smoke coming from his homestead and rushes back home. While on his way home another lightning bolt strikes the beer parlour where he had gone for a drink. Again, nobody is killed but people are terrified. Before he reaches his home, yet another lightning bolt strikes Achebe and kills him.

What is your view about what you have just read? Why was lightning attracted to Achebe or why did Achebe attract lightning?

Step 2: Determine the intended learning outcomes

In teaching electrostatics, conceptual resources (e.g., charges, conductors, etc.) for studying lighting in terms of science and indigenous knowledge may be used to promote space for argumentation as preamble. An appropriate argumentation scheme that facilitates the integration, or a fair comparison, of science and indigenous knowledge (i.e. caters for both logical and non-logical aspects of argument) should be used. CAT is very useful in achieving this.

Step 3: Differentiate between the intended outcomes and learners' conceptions (including nature of conflicts)

In predicting how the learners negotiate the explanation of lightning as the cause of Achebe's death, and to ascertaining whether they develop clear argumentation division lines along different worldviews (i.e. linear causal reasoning) (Driver, Guesne, & Tiberghien, 1985), the facilitator (teacher) can play devil's advocate (e.g. posing provocative questions- how do you know?) in the argument with learners. However, if TWO systems of thought emerge from learners' responses, which claim to be scientific and indigenous, such responses should be treated side by side. This learning outcome involves another type of reasoning, which Perkins and Grotzer (2005) call relational causal reasoning. In other words, it is the relation between the two worldviews rather than just the one that determines whether it is the scientific or indigenous knowledge that explains what killed Achebe. The desire to compare, to measure and to categorise in terms of replacing one with the other, or determining which is better and more acceptable, can be probed and identified using CAT categories (i.e. dominant, suppressing, assimilating, emergent and equipollent). Assisting learners in their small or individual groups to see, accept and correct the mistakes in their arguments should be done with tact and good humour when critiquing them. It is important for the learners to realise that human beings use or rely on different worldviews to explain their experiences. The teacher should call learners' attention to instances where indigenous knowledge and scientific arguments agree/disagree. Make learners aware of how their science or indigenous knowledge influences their argumentation skills.

The teacher would then mediate the whole class argumentation process (as deemed necessary in the DAIM setting) to identify counter-claims and rebuttals using the five categories of CAT. The class agrees on the various levels of argumentation before reaching a consensus on whether it is the scientific or indigenous knowledge that explains how the lighting killed Achebe.

Another important conclusion is that CAT has also been used as an analytical tool for determining perceptual shifts that tend to occur in argumentation, e.g. shifting from a scientifically dominant stance to a traditional stance as the contexts of the arguers change. In addition, CAT may to a large extent provide learners with opportunities to examine competing ideas, evaluate the evidence that does or does not support each perspective, and construct arguments justifying the case for one idea or another. In line with CAT, it can also be said that DAIM may not only provide opportunities for learners in small groups to evaluate alternative ideas of their peers, but may also encourage them to use evidence to distinguish between these ideas in a more rational way (Linn & Eylon, 2006; White & Gunstone, 1992). The implication here is that DAIM is a viable method of teaching science, which can be used fruitfully in other subject areas as well.

Conclusion

To make the teaching of science through argumentation as a teaching method feasible in science classroom, I would recommend that teachers be equipped with pedagogical skills more compatible with cross-disciplinary teaching approaches that provide ample learning opportunities for learners. The role of the teacher in this regard is that of a facilitator rather than a disseminator of a decontextualised knowledge (Ogunniyi, 2007a; Shumba, 1999). If teachers are to implement the argumentation scheme successfully, they should be made aware of alternative worldview perspectives, identify their merits and demerits, the similarities and the differences, so that they can use the understanding gained to make wise decisions in their classrooms as well as their daily lives (Ogunniyi, 2011). It is therefore imperative to find ways of helping the teachers to improve their organisation of pedagogical and epistemological knowledge of science in order to give them room for engaging learners in meaningful scientific experiences in their classrooms. It is when we, as teacher educators, figure out how we can help teachers to mediate in disagreements with reason that argumentation studies will extend the historical modes of argument embodied for centuries in Plato and Aristotle's times.

References

Asterhan, C., & Schwarz, B. (2007). The effect of monological and dialogical argumentation on concept learning in evolution theory. *Journal of Education Psychology, 99*(3), 626–639.

Ausubel, D. (1968). *Educational psychology: A cognitive view.* New York, NY: Holt, Rinehart, Winston.

Berland, L. K., & Reiser, B. R. (2008). Making sense of argumentation and explanation. *Science Education, 93*, 26–55.

Chinn, C. A., O'Donnell, A. M., & Jinks, T. S. (2000). The structure of discourse in collaborative learning. *The Journal of Experimental Education, 69*(1), 77–97.

Clark, D., Sampson, V., Weinberger, A., & Erkens, G. (2007). Analytic frameworks for assessing dialogic argumentation in online learning environments. *Educational Psychology Review, 19*(3), 343–374.

Department of Basic Education. (2011). *Curriculum and assessment policy statement (CAPS) Physical Sciences*: Final draft. Pretoria: Republic of South Africa.

Department of Education. (2016). *National diagnostic report on learner performance in Physical Sciences grade 12*. Pretoria: Republic of South Africa.

Diwu, C., & Ogunniyi, M.B. (2012). Dialogical argumentation instruction as a catalytic agent for integrating science with Indigenous knowledge systems. *African Journal of Research in Mathematics, Science and Technology Education, 16*(3), 333–347.

Driver, R., Guesne, E., & Tiberghien, A. (1985). *Children's ideas in science*. Milton Keynes: Open University Press.

Driver, R., Newton, P., & Osborne, J. (2000). Establishing the norms of argumentation in classrooms. *Science Education, 84*(3), 287–312.

Encyclopedia Britannica. (2008). *Artificial intelligence*. Retrieved from http://search.eb.com/eb/article-219080

Erduran, S., Simon, S., & Osborne, J. (2004). TAPping into argumentation: Developments in the application of Toulmin's Argument Pattern for studying science discourse. *Science Education, 88*(6), 915–933.

Erduran, S. (2007). Methodological foundations in the study of argumentation in science classrooms. In S. Erduran & M. P. Jiménez-Aleixandre (Eds.), *Argumentation in Science Education, 35*, 47–69. Springer Netherlands.

Heller, P., Keith, R., & Anderson, S. (1992). Teaching problem solving through cooperative grouping. *American Association of Physics Teachers, 60*(7), 627–636.

Hewson, P. W. (1996). Teaching for conceptual change. In D. F. Treagust, R. Duit, & B. J. Fraser (Eds.), *Improving teaching and learning in science and mathematics* (pp. 131–140). New York, NY: Teachers College Press.

Jimenez-Aleixandre, M. P., & Erduran, S. (2008). Argumentation in science education: An overview. In S. Erduran & M. P. Jimenez-Aleixandre (Eds.), *Argumentation in science education: Perspectives from classroom-based research* (pp. 3–28). Netherlands: Springer.

Jimenez-Aleixandre, M., Rodrigues, A., & Duschl, R. (2000). "Doing the lesson" or "doing science": Argument in high school genetics. *Science Education, 84*, 757–792.

Kelly, G., Druker, S., & Chen, C. (1998). Students' reasoning about electricity: Combining performance assessments with argumentation analysis. *International Journal for Science Education, 20*(7), 849–871.

Kelly, G., & Takao, A. (2002). Epistemic levels in argument: An analysis of university oceanography students' use of evidence in writing. *Science Education, 86*, 314–342.

Kuhn, D. (2010). Teaching and learning science as argument. *Science Education, 94*(5), 810–824. doi:10.1002/sce.20395

Kuhn, D., & Udell, W. (2003). The development of argument skills. *Child Development, 74*(5), 1245–1260.

Lave, J., & Wenger, E. (1991). *Situated learning: Legitimate peripheral participation.* New York, NY: Cambridge University Press.

Lawson, A. E. (2004). The nature and development of scientific reasoning: A synthetic view. *International Journal of Science and Mathematics Education, 2*(3), 307–338.

Lehrer, R., & Schauble, L. (2006). Scientific thinking and science literacy: Supporting development in learning in contexts. In W. Damon, R. Lerner, K. A. Renninger, & E. Sigel (Eds.), *Handbook of child psychology* (6th ed., pp. 153–196). Hoboken, NJ: Wiley.

Lemke, J. (1990). *Talking science: Language, learning and values.* Norwood, NJ: Ablex Publishing.

Linn, M., & Eylon, B. (2006). Science education: Integrating views of learning and instruction. In P. Alexander & P. H. Winne (Eds.), *Handbook of educational psychology* (pp. 511–544). Mahwah, NJ: Erlbaum.

Ogunniyi, M. B. (1997). Science education in a multi-cultural South Africa. In M. Ogawa (Ed.), *Science education and traditional cosmology: Report of an international research programme (Joint Research) on the effects of traditional culture on science education* (pp. 84–95). Mito: University Press.

Ogunniyi, M. B. (2004). The challenge of preparing and equipping science teachers in higher education to integrate scientific and indigenous knowledge systems for learners. *South Africa Journal of Higher Education, 18*(3), 289–304.

Ogunniyi, M. B. (2007a). Teachers' stances and practical arguments regarding a science-indigenous knowledge curriculum, part 1. *International Journal of Science Education, 29*(8), 963–986.

Ogunniyi, M. B. (2007b). Teachers' stances and practical arguments regarding a science-indigenous knowledge curriculum, part 2. *International Journal of Science Education, 29*(10), 1189–1207.

Ogunniyi, M. B. (2009, October 29–31). *Implementing a science–indigenous knowledge curriculum, The Western Cape experience.* Paper presented at the Second National Workshop on Science and Indigenous Knowledge System, University of the Western Cape.

Ogunniyi, M. B. (2011). The context of training teachers to implement a socially relevant science education in Africa. *African Journal of Research in Mathematics, Science and Technology Education, 15*(3), 98–121.

Osborne, J., & Patterson, A. (2011). Scientific argument and explanation: A necessary distinction? *Science Education, 95*(4), 627–638.

Perkins, D. N., & Grotzer, T. A. (2005). Dimensions of causal understanding: The role of complex causal models in students' understanding of science. *Studies in Science Education, 41*(1), 117–166.

Popper, K. (1968). *The logic of scientific discovery.* New York, NY: Harper & Row Publishers, Inc.

Shumba, O. (1999). Relationship between secondary science teacher's orientation to traditional culture and beliefs concerning science instructional ideology. *Journal of Research in Science Teaching, 36*(3), 333–355.

Simon, S., Erduran, S., & Osborne, J. (2006). Learning to teach argumentation: Research and development in the science classroom. *International Journal of Science Education, 28*(2–3), 235–260.

Toulmin, S. E. (2003). *The uses of argument.* (First published 1958) Cambridge, England: Cambridge University Press.

Van Eemeren, F. H., Grootendorst, R., & Kruiger, T. (1987). *Handbook of argumentation theory.* Dordrecht: Foris Publications.

Wandersee, J. (1985). Can the history of science help science educators anticipate students' misconceptions? *Journal of Research in Science Teaching, 23*(7), 581–597.

Wenger, E. (1998). Identity in practice. In E. Wenger (Ed.), *Communities of practice: Learning, meaning, and identity* (pp. 149–163). Cambridge, UK: Cambridge University Press.

White, R., & Gunstone, R. (1992). Probing understanding. London: Falmer Press. Wilkerson-Jerde, M., & Wilensky, U. (2011) "How do mathematicians learn math?: Resources and acts for constructing and understanding mathematics," *Educational Studies in Mathematics,* 21–43.doi:10.1007/s10649-011-9306-5.

Zohar, A., & Nemet, F. (2002). Fostering students' knowledge and argumentation skills through dilemmas in human genetics. *Journal of Research in Science Teaching, 39*(1), 35–62.

The Effect of a Professional Development Programme on Science Teachers' Instructional Practices

Introduction

As the aim of this study was to measure the effect of a professional development programme (PDP) on science teachers' pedagogical practices, I want to start by drawing attention to Lee Shulman's (1987) statement that 'comprehended ideas must be transformed in some manner if they are to be taught' (p. 16). Shulman (1987) believes that if teachers want their learners to learn, they must deliver the content in a pedagogically powerful way. To him teaching is principally the responsibility of the teacher to engage his or her learners in the learning process in a meaningful way, which he refers to as a 'pedagogy of engagement'. In South African schools researchers in national reports and polemics often highlight the point that Physical Science teachers do not pedagogically engage their learners in a powerful way. For example, a report by the Foundation for Research and Development (FRD) (1993) revealed that in the mid-1980s only 16% of black learners passed Physical Science compared to 80% of mixed race, Indian and white learners. They attribute the poor performances of the learners to poor teaching. Furthermore, this report makes the point that less than 5% of teachers in black schools are qualified to teach Physical Science and that the majority of the teachers are stuck in traditional pedagogies. By traditional pedagogies I mean teaching approaches in which teachers hand to their learners predesigned, ready-made knowledge that is inappropriate, meaningless, void of practical work and disengaging. Two decades later Basson and Kriek's (2012) study revealed that nothing had changed as teachers still struggle with content and therefore lack effective pedagogies.

To Shulman (1986) powerful pedagogies are critical for teachers if we want our learners to develop better life-long learning skills and life-deepening attitudes that promote greater learner understanding and higher achievement than is possible with more traditional pedagogies, which promote passive forms of learning. What is needed in South Africa are instructional approaches that engage learners critically and meaningfully—what Shulman (2005) refers to as pedagogies of engagement. According to Lee and Kraphl (2002), one of the best ways to introduce teachers to such new instructional practices is through a PDP, as this may help teachers significantly to gain the technical skill, knowledge and experience necessary to engage in reform-orientated teaching. Dass (2001) points out that this is possible because PDP is designed in such a way that the needs of the participants are met. Furthermore, Klein (2001) asserts that PDP activities or workshops must be designed in such a way that they are aligned with the legislated curriculum. Hence the aim of this PDP was to introduce the teachers to a phenomenological instructional method that promotes knowledge by acquaintance, which involves learner lived-experience centred on and aligned with the CAPS. The idea in the workshops was to make the teachers aware of the knowledge, ideas and scientific understanding of learners through learners' personal engagement with nature. We wanted the teachers to be cognisant of the learners' lived world experiences and to connect with their ideas, beliefs and understanding of nature in the planning of their lessons.

Although the current South African CAPS states that science teachers must adopt innovative teaching strategies in the delivery of their content, researchers reported that many teachers did not receive the necessary training to implement the suggested instructional approaches (Koopman, 2013; Koopman, Le Grange & de Mink 2016). The lack of experience makes it very difficult for teachers to change their traditional pedagogies based on passive rote learning in which the minds of the learners are seen as blank slates to write knowledge on.

This brings me to the aim of this study, which was guided by the following research question: What is the effect of PDP on Physical Science teachers' instructional practices? As discussed in the above sections, the goal of this PDP, which took the form of workshops, was to engage the teachers in a phenomenological instructional method and to evaluate the effectiveness of such programmes. I will now provide a succinct overview of the state of Physical Science teaching in South Africa as a rationale for the study.

Research Related to the Study

Many previous studies on teachers' instructional practices report on science teachers' 'failure' to 'excite' and 'engage' learners effectively in the science classroom. According to Kazeni and Onwu (2013), the reasons why teachers continue to fail to excite and attract learners to engage with the subject meaningfully are: (i) the subject is taught as a rote listing of facts that learners find irrelevant and meaningless to their personal lived world; (ii) teachers fail to raise their learners' awareness of the social impact of science, with particular reference to the rewards embedded in the subject, such as science's potential beneficial effect on personal development; (iii) science teachers lack effective teaching approaches that link the content of the subject to the day-to-day lived experiences of learners; this is likely to further obscure and diminish the relevance of the subject in their lives. Other studies, both local and abroad, further highlight how science teachers are stuck in traditional, behaviouristic instructional pedagogies that are dull and boring. (For a full account, see Bennett & Holman, 2002; Centre for Development and Enterprise [CDE], 2010; Kazeni & Onwu, 2013; Naidoo & Lewin, 1998; Reddy, 2006.) This happens, as Dube and Lubben (2011) point out, because of the inability of science teachers to develop teaching approaches that engage learners actively by linking propositional knowledge (factual) with knowledge by acquaintance (personal lived experience). When teachers bring these two knowledge types together, they give legitimacy to their learners' critical thinking, enabling them to see the value of science in their everyday experiences in the real world.

One of the aims of the shift in 1997 from the apartheid curriculum (NATED 550 Report, n.d.) to the various post-apartheid curricula, C2005, RNCS, NCS and CAPS (for full details see Department of Basic Education [DBE], 2008, 2011a, 2011b; Department of Education [DoE], 1999) was to break from the sterility of canonical science, so that teachers can change their traditional pedagogies and adopt innovative teaching strategies that actively involve the learners in the teaching and learning process. CAPS, which is currently the only legislated curriculum framework, is a modification of its predecessor—the NCS. CAPS for Physical Science has three aims, which promote three knowledge types: (i) *Specific aim 1*: investigating phenomena (*knowledge by acquaintance*); (ii) *Specific aim 2*: construction of scientific knowledge (*propositional knowledge* or *knowing that*) and (iii) *Specific aim 3*: appreciating and understanding the link between science, technology and society *(knowledge of knowing how)*. These three specific aims and the knowledge discourse dominate the epistemic ascent of the new Physical Science curriculum. It follows, then, that the subject involves significant practical abilities that are dependent on experience and investigation rather than revelation or calculations. When teachers promote these knowledge types, they assist their learners in understanding the procedural knowledge and scientific skills that learners can internalise in their everyday engagement with the material world. Furthermore, these knowledge types also assist learners to solve personal scientific problems and to make responsible socio-scientific decisions. While many design studies and reports focus on learners, sociocultural issues, systemic challenges, scarce resources, international benchmark tests, and the curriculum, there is a concern that they devote little attention to the pedagogical practices of science teachers and how to assist teachers in this regard.

Theoretical Framework

Phenomenology as an Instructional Approach

In his book entitled *Science Education and Curriculum in South Africa*, Koopman (2017) asks the question whether a phenomenological instructional approach to science teaching might improve the quality of teaching and learning in South Africa. He critically discusses and compares the various didactical approaches of behaviourism, cognitionism and phenomenology, their underpinning philosophies, rationale and axiological importance, and argues compellingly for a shift towards adopting phenomenological pedagogies. He avers that phenomenology as a teaching method is not interested in knowledge about *'knowing that'* or *'knowing about'* science, but about *doing science* like a real scientist at work. He argues that doing science has the potential to foster creativity, open-mindedness and critical thinking. His contention is that, when teachers focus on the learners' lived experience, coupled with the use of their senses, learners could be led to new insights whilst dispelling deeply rooted false beliefs and perceptions. Koopman is of the opinion that the most constructive response to this *'barrenness'* of 'good' teaching lies in the pedagogy of phenomenology. International phenomenological scholars such as Ted Aoki (1992), Max van Manen (1990), William Pinar and Reynolds (1992) and various others argue that phenomenology is not only a pedagogy of 'active engagement', but brings three essential features into the classroom—the act of caring, thoughtfulness and a sensitivity to the personal needs of the learners.

Bo Dahlin (2001) had already, like Koopman (2017), averred that phenomenology as a teaching method brings man and nature closer. This approach to teaching does not only teach the learner to appreciate the beauty of nature but also to understand it. When teachers adopt a phenomenological method, in Dahlin's (2001) view, they bring the learner to a state of awareness in which he or she learns to listen when nature speaks through the gestures it makes in its different forms—colours, smells and tastes. In other words, the teachers do not only break down the thick walls that separate their learners from nature, but teach them to listen to all that nature has to say. Aoki (1992) argues that phenomenology helps us to focus on our *'isness'* (p. 190), which calls for a breaking away from the various orientations in our everyday life that blind us to the world around us. To him, phenomenology helps us to break through all the layers outside of us in an endeavour to promote a closer connection with our earthly dwelling. Van Manen (1990) shares similar sentiments and states that in the area of pedagogy, phenomenology guides teachers on how to act tactfully, on the basis of a carefully cultivated thoughtfulness.

The above statements and descriptions suggest that if teachers embrace the lived world of their learners and are aware of what it is that excites them, they will learn what and how to teach. It is important to note here that children are born natural scientists with an inborn sense of curiosity. In other words, science is part of a child's natural world and therefore becomes a source of learning in the Physical Science classroom. Thus understood, when teachers pay careful attention to the lived world experiences of their learners, they will know what it is that excites them and at the same time know how to teach the child. This also forms the essence of a phenomenological pedagogical approach, because a child is the product of a network of relationships with the material world which he or she chooses and to which he or she is chosen. This raises the question: What types of knowledge must teachers promote in a phenomenological approach? To answer this question I will provide a brief overview of the kind of propositional knowledge that has dominated behaviouristic and cognitive learning for many decades in South Africa, followed by a discussion of inculcating knowledge by acquaintance, an approach which teachers can use in the teaching and learning of science.

Propositional Knowledge

Research over the last few decades has revealed that teachers view an academic subject such as Physical Science as being mainly composed of propositional knowledge rather than practical knowledge. Science as a discipline is a practice that entails specific procedures for knowledge production; this knowledge is validated and established as truth by means of practical evidence that is closely connected to a conceptual structure. Mastering this structure requires not so much propositional knowledge, but rather practical knowledge and knowledge by acquaintance.

According to Brandom (2000), propositional knowledge consists of two parts, namely concepts that are linked to a proposition. He asserts that a grasp of the concepts is embodied in a proposition. This means concepts only make sense if a person has an understanding of the inferences that can be carried out with the propositions to which that concept is linked (p. 28). In Brandom's view (Brandom, 2000), a person cannot grasp the content of a proposition unless that person can understand the inferences that can be drawn from it or to it. For example, a conceptual understanding of an electron orbiting the atom requires an understanding of the spatial dimensions of the atom and the inferences drawn about the various particles making up the atom. In other words, an understanding of the concept 'electron' is not only based on the knowledge we have acquired about electrons, but also on the belief that electrons exist. On the basis of this premise Winch (2013) argues that the knowledge a person holds—for example, about an electron in this case—does not mean it can be better understood by someone else, seeing that no one has ever experienced an electron in the real world. Instead the only knowledge that individuals can have about the electron is its practical connectedness with other propositions in the atom.

This kind of propositional knowledge emerges when teachers shy away from the personal lived and complex world of learners. Aoki (1992) asserts that when teachers focus on propositional knowledge, their teaching approaches are situated in a black box which in the process reduces the learners to passive vessels. This is because the focus of the content is on embodied propositions drawn from inferences articulated from the lived world of teachers. This implies that teachers view their role as that of line managers who package knowledge for learners (ready-made knowledge)—likening the school to a factory or a knowledge industry. Their teaching approaches resemble the 'factory' model of schooling inspired by Frank Tyler (1949) and his emphasis on industrial design, which was the blueprint for teachers on how to teach for decades.

Knowledge by Acquaintance

According to Winch (2013, p. 129), when a person 'hears a symphony, smells a flower and tastes a fruit' he or she is said to be acquainted with them. He asserts knowledge by acquaintance can only be experienced by living in the moment and connecting with objects with the senses. He points out that it is also possible to give a description of a landscape without taking into consideration what the observer embrace, such as the colours, the vista and the atmospheric condition, the various smells and sounds. From this perspective knowledge by acquaintance cannot be fully supplanted by a full and accurate description. Such knowledge (or sense-experiences) cannot be theorised or presented accurately. Our senses are biological instruments through which we communicate with the world and vice versa. However, our senses are feeble—but through scientific technology humans have learned to extend the range and power of their senses. For example, hundreds of years ago our eyesight was limited to the colours of the rainbow. Back then the universe not only communicated with us through that tiny slice of the electromagnetic spectrum, but most of our discoveries were made through them. Today teachers ignore the skill and power of observation, taste, smell, hearing and so forth and as a result these things are given little attention in the science classrooms. Instead of helping learners to develop their own rulebook for how the world works with the use of their senses, they are given ready-made knowledge and rules described in textbooks. Therefore it goes without saying that knowledge by acquaintance can have a positive impact on a learner's understanding of the scientific world.

Personal lived experience or knowledge by acquaintance is experiential and has experiential orientations towards nature. These experiences reduce the external world to imprints on the mind. For this reason Locke (2009) believes that man's senses convey deep into the mind several distinctive perceptions and properties of things. In his view, it is only through sensations that one arrives at a full understanding of one's environment, the world or things. He contends that a person has no knowledge if he or she does not have experience which is stimulated by his or her senses. Therefore, when an individual becomes acquainted with a phenomenon, according to Heidegger (2002), his/her perceptive knowledge collapses in favour of knowledge by acquaintance. In other words, it is only through lived experience that a learner might discover truth, or what works and what does not work in real life. This so-called discovery (or truth about phenomena) might lead to a deeper and richer understanding of a concept or propositional knowledge, as the individual can delve deeper into the phenomena through insightful questioning. Knowledge by acquaintance can lead to people asking more and also more appropriate questions, which in turn have the potential to add more value to their knowledge. For example, Galileo arrived at his conclusions about gravity, and Kriek and Watson solved the DNA problem in 1958, by asking *the right questions*. This is because the environment of the scientist is always changing. Therefore knowledge by acquaintance provides the platform for scientific revolutions. One of the attributes of welcoming lived experience into the science classroom is that learners are free to ask insightful questions and at the same time also learn the value of good questioning.

Halling (2002, p. 6) argues that phenomenological pedagogies (centred on lived experience) might lead to verifying and validating phenomena through our own experience of what we had previously only heard of or seen outside of ourselves. This verification and validation of knowledge through experience represent an individual's dialogue with the world through which he or she establishes a true belief in a phenomenon. This validation of knowledge through experience, according to Heidegger (1967), is what phenomenology is all about. Heidegger used the term 'ontology' to describe phenomenology because ontology is a study of the modes of being in the world. This is why in a phenomenological paradigm, asking questions guides an individual in his or her spiritual quest for knowledge about the nature of objects in order to ascribe meaning to them.

Aoki (1992) points out that teachers need to open and enter this black box of teaching. By entering the black box and separating the instrumental and technical from the axiological, he argues, we can transport our learners to higher forms of learning. In other words, Aoki views effective science teaching as facilitating the flow of knowledge from that with which a learner is acquainted to propositional knowledge, and not vice versa. This means the learner becomes a co-constructor with the teacher in the knowledge acquisition process, while in the process understanding how knowledge generation in the scientific world takes place. This is because nature becomes the learner's laboratory, while he or she searches for meaning and understanding of all the physical and chemical processes around him or her. This not only leads to higher-order critical thinking, but makes the complex simple and adds value to the concept of learning as the learner can relate at his or her level to the content he or she is taught.

Instructional Context

This study stems from a collaborative project between a consortium of academics from three South African universities and the Department of Basic Education (DBE) in Limpopo province. Before the commencement of the project, the DBE had expressed concern about the inadequate instructional practices of their Physical Science teachers. Based on this point of departure, one of the goals of this project is to assist science teachers in improving their pedagogical knowledge (PK). The research team engaged the teachers in a week-long workshop (eight hours per day) once a year for three years. In these workshops each facilitator presented at least one topic per day for two hours on phenomenology as a teaching methodology. In every topic we addressed the merits and principles of phenomenology by discussing its importance and value in a teaching methodology. This instructional method was also modelled to the teachers in some of the sessions. Figure 6.1 gives a diagrammatical representation of the sequencing of events in a phenomenological paradigm.

Figure 6.1: Sequencing of events in a phenomenological classroom.

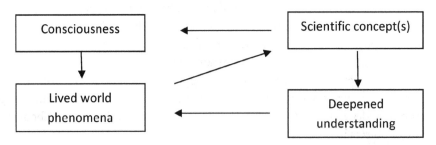

Adapted from: *6th IOSTE Symposium for Central and Eastern Europe,* Østergaard, Dahlin, and Hugo (n.d.).

Consciousness represents the learner's active thinking as derived from experience. According to Østergaard *et al.* (n.d.), the teacher should:

- allow the learner to draw on his or her own experience to create a phenomenon for the content;
- link the phenomenon to the scientific explanation by using the learner's own life world as point of reference;
- use the scientific explanations and/or findings or descriptions to deepen the learner's understanding of the phenomenon;
- expand on the science to deepen the learner's understanding of the phenomenon.

During these workshops the teachers were also encouraged to share their ideas of how they think it best to deliver the content.

On day three of the first workshop in year one the teachers divided themselves into five groups of three teachers. No criteria were given to them for this division into groups. Each group had to plan and present a lesson of their choice from the Grade 10, 11 or 12 syllabuses to their peers (who pretended to be Grade 10 learners). The brief given to the teachers was:

- Select a chemistry topic from the Grade 10, 11 or 12 syllabus and plan, design and present a 30-minute lesson;
- The lesson had to be relevant to and captivating for the learners;
- The content must be challenging and not oversimplified.

The teachers were given the brief at the close of the previous day so that they would have enough time to prepare. They were also encouraged to exchange ideas in the planning of their lessons. The lessons were presented in the conference venue and each group member had to make a contribution. The topic the groups chose in year one was repeated and expanded on in the following year to assess whether the comments made in the previous workshop were included in the lessons. The presentations in the second year took place on day two of the workshops, without giving the teachers detailed instructions. Instead, their brief was that they had to present a lesson on the same topic. A total of 10 lessons (five lessons in year one, and five more in year two of the workshop) were video-recorded, followed by post-lesson interviews/discussions.

Methodology

In this study both quantitative and qualitative data-construction methods were used. On the quantitative side an 'interrupted group' time series design was used, given that the time interval between the two stages of data collection and intervention was approximately a year long. The reason for using this design was because the researcher and the participants only had a week of contact because of financial constraints. The participants resided in a different province from the researcher. Another reason was that the focus of this study was to measure the effect of the PDP over a period of two years. This was done to see if there were any changes demonstrated in the teachers' instructional practices. In this case no baseline intervention was undertaken and the focus was only on the effect of the PDP. Qualitative data were constructed from post-lesson discussions and field notes that were used to augment the quantitative results.

The participants were 15 'Lead' teachers and two science education specialists. Lead teachers were chosen in cluster meetings and they are normally more experienced than others. They are considered competent in their subject field and have a record of excellent results. The teachers represented fifteen different districts in Thohoyandou. After the participating teachers had been empowered with instructional strategies, they were required to share their experiences and expertise with others and mentor novice teachers in their districts. The teachers were well experienced and all of them taught Natural Science and/or Physical Science at their respective schools (for full details, see Table 6.1).

Table 6.1: Background information of the participants

Teacher	Gender	Race	Teaching experience (years)	Post level at school	Qualifications	Subject major
1	Male	African	20	CS 1 –teacher	BSc	Chemistry, Physics, Mathematics
2	Male	African	22	CS 2 –HOD	3–year diploma	Physical Science, Mathematics
3	Male	African	31	CS 1 –teacher	BEd (Hons)	Physical Science, Life Science
4	Male	African	30	CS 1 –teacher	BEd (Hons)	Physical Science, Life Science
5	Female	African	10	CS 2 –HOD	STP, ACE	Physical Science, Mathematics
6	Female	African	8	CS 1 –teacher	STD	Agricultural Science Life Science
7	Male	African	14	CS 2 –HOD	BSc (Ed), BEd (Hons)	Physical Science, Mathematics

Teacher	Gender	Race	Teaching experience (years)	Post level at school	Qualifications	Subject major
8	Male	African	10	CS 1 –teacher	1 – BA (Ed), BEd	History
9	Female	African	2	CS 1 –teacher	BSc, PGCE	Chemistry, Physics
10	Male	African	21	CS 1 –teacher	STD, ACE	Agricultural Science, Geography
11	Male	African	18	CS 1 –teacher	STD, ACE	Natural Science, Mathematics
12	Female	African	21	Curriculum advisor	BSc, BEd (Hons)	Physical Science, Mathematics
13	Female	African	15	CS 1 –teacher	STP, ACE	Physical Science, Mathematics
14	Male	African	14	CS 3 –deputy principal	BSc, PGCE	Physical Science, Life Science
15	Male	African	25	Curriculum advisor	BSc, PGCE	Physical Science, Life Science

Source: Author.

The research team (or facilitators) presenting the workshops were a consortium of academics from three different universities, specialising in Chemistry, Physics, Zoology and Botany respectively. The author of this article is a Chemistry specialist. All the facilitators were well experienced in both school and university teaching, with not less than 10 years' university teaching experience. All three facilitators had doctoral degrees in their respective disciplines.

Data Sources

The data sources used for this study were 10 video recordings and post-lesson interviews or discussions. During each presentation a video was made of the whole group and each lesson was analysed to determine whether or not the presenters had used the new methodology. The responses were probed where necessary to ensure that there were no misconceptions or misunderstandings as to why they did what they did in their presentations.

Data Analysis

Each lesson was video recorded and analysed with reference to a phenomenological observation protocol (POP), which is a reconstruction of the Reform Teacher Observation Protocol (RTOP) (Piburn et al., 2000). POP was validated by five experts in the field of phenomenology using Spearman ranking correlation (. The value was calculated using the formula after all the experts had ranked each item. The Spearman ranking value is within the acceptable range of This implies that overall there was a strong positive agreement between the rankings of the five experts with respect to the validity of the instrument and that the instrument carefully covered all the items it intended to measure.

Table 6.2: The POP items related to inquiry orientation

Subset	POP item		
Lesson design	1.	1.	The lesson was learner-centred.
	2.	2.	Personal lived experiences were favoured.
	3.	3.	Scientific knowledge was used to deepen the learners' understanding of everyday phenomena.
Knowledge by acquaintance	1.	4.	The teacher used a variety of scenarios to capture the learners interest and encourage debate.
	2.	5.	Learners were actively engaged in thought-provoking activities which were relevant to their lived world.
	3.	6.	The teacher invited learners to share ideas about the topics.
	4.	7.	The teacher used questions such as 'Did you ever see, touch, smell or hear of these things we are discussing?'
	5.	8.	The teacher gave prominence to the use of the senses in the lesson.
Propositional knowledge	1.	9.	The lesson was more focused on factual knowledge.
	2.	10.	Very little attention was given to the learners' personal lived experiences.
	3.	11.	Learners could not connect with the content as it disregarded their experiences.
Classroom culture	1.	12.	Learners participated actively in the lesson.
	2.	13.	They were allowed to exchange ideas with one another and in the group.

Adapted from: Piburn et al. (2000).

Table 6.2 highlights the 13 items used to evaluate each lesson. The core of POP is based on four subsets: (i) design and implementation; (ii) knowledge by acquaintance; (iii) propositional knowledge and (iv) classroom culture. Each subset was further divided into different items, such as whether the teacher designed a learner experience-centred lesson and whether the knowledge disseminated in the science classroom leaned more towards knowledge by acquaintance or propositional in nature. Focusing first on knowledge by acquaintance, and then linking it with propositional knowledge, can lead to deeper and richer understandings of the scientific world. The main idea was to connect the lived world experiences of learners with objects in their world to the scientific explanations.

POP was used in conjunction with a Likert scale that ranged from 0 to 4 (0: not observed to 4: very observed). Each of the ten lessons was viewed many times before ranking each item under the various subsets. When the researcher was confident of the ranking of each lesson (0–4), a decision was made. POP was further augmented with the post-lesson interview notes to develop an understanding of whether the teachers had adopted a phenomenological method. After analysing the videos and drafting my main findings, I invited a colleague, an expert in phenomenology who was unfamiliar with the data, to watch the videos with me in order to validate my findings. We looked at the videos together and encapsulated the different methodologies used and discussed my main findings. This led to a refinement in the descriptions of the distinctions between the different levels. We also discussed the verbal explanations of the teachers for each lesson and how they described concepts; the accuracy of their descriptions were noted during the analysis. All disagreements between us were revised by consensus. Next I present the findings of the five lessons.

Findings

A graphical representation of the data in Table 6.3 was constructed in Figure 6.1. While Table 6.3 shows all the quantitative data of the teachers as completed on the POP, Figure 6.1 is a graphical representation of a divergent Likert scale. This means the data of each lesson were separated and presented separately on the same graph. Table 6.3 shows that the scores increased for items 1 and 2 for subsets 1, 2 and 4, and decreased for subset 3. The first lessons presented by groups 1, 2 and 5 leaned more towards traditional pedagogies, while lesson 1 of group 3 and 4 adopted more of a phenomenological approach from the outset. The findings also show that there was a consistent increase in the scores of all the groups from lesson 1 to lesson 2. This is because in year 2 of the programme the teachers became more familiar with phenomenology as a teaching methodology and therefore found it easier to make the transition in favour of the phenomenological approaches. Thus, all the teachers paid more careful attention to (or at least considered) the lived experiences of the learners by focusing more on the macroscopic properties of chemical substances, while little attention was given to the symbolic and sub-microscopic properties. The macroscopic properties in science focus more on the external conditions of objects that people can see, touch, taste and smell, while the sub-microscopic properties refer more to the materialistic composition of substances. This means the smaller or microscopic elements of matter that cannot be seen with the naked eye. Focusing more on the macroscopic properties of substances also implies that the teachers' pedagogies became more learner-centred as opposed to the traditional more teacher-centred pedagogies with their focus on propositional knowledge.

All the groups in their first lesson struggled to make the connection with the real-life events of the learners and in the process paid very little attention to knowledge by acquaintance. However, as the PDP progressed to assist the teachers in understanding the implementation of phenomenological principles in the second lesson, small strides were made by groups 1, 2 and 5, while tremendous progress was made by groups 3 and 4. Next I discuss the types of knowledge the teachers promoted in their presentations.

Table 6.3: POP scores of the teacher over a period of two years

Subset	POP item	Grup 1		Grup 2		Grup 3		Grup 4		Grup 5	
		01234		01234		01234		01234		01234	
Lesson design	1. The lesson was learner-centred.	1	2	0	2	2	3	2	3	0	3
	2. Personal lived experiences were favoured.	1	2	1	2	2	3	2	2	0	2
	3. Scientific knowledge was used to deepen the learners' understanding of everyday phenome.	0	2	1	(2)	1	2	2	3	(0)	2
Knowledge by acquaintance	4. The teacher used a variety of scenarios to captivate the interest and encourage debate.	1	2	1	2	2	2	2	3	0	2
	5. Students were actively engaged in thought-provoking activities and discussions that were relevant to the learners' lived world.	1	2	(1)	2	2	3	(1)	3	0	2
	6. The teacher invited learners to share their ideas about the topics.	1	2	1	(2)	1	2	1	3	0	2
	7. The teacher used questions like "Did you ever see, touch, smell or hear any of these things we are discussing?"	0	1	0	1	0	1	1	2	0	1
	8. The teacher gave prominence to the use of the senses in the lesson.	(0)	1	0	1	0	(1)	1	2	0	1

Subset	POP item	Grup 1		Grup 2		Grup 3		Grup 4		Grup 5	
Pro-positional knowledge	9. The lessom was more focused on factural knowledge.	3	1	4	1	3	2	2	2	3	2
	10. Very little attention was paid to the learners' personal lived experiences.	3	2	3	2	3	(2)	(2)	2	(3)	2
	11. Learners could not connect with the content as it disregarded their experiences.	3	2	2	2	3	2	2	2	3	2
Classroom culture	12. Learners participated actively in the leasson.	3	2	3	2	3	2	2	2	(3)	1
	13. They were allowed to exchange ideas with one another.	1	2	1	2	(0)	1	1	2	2	2

Y – Me
(Y) – Hinted by me
Source: Author.

Figure 6.2: Graphical representation of POP scores of teachers' lessons

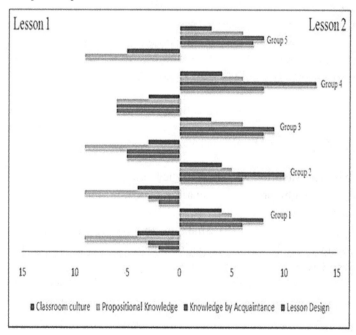

Source: Author.

The findings of Figure 6.2 show that in lesson 1 all the groups paid more attention to propositional knowledge, while less attention was given to knowledge by acquaintance. This approach decreased drastically in lesson 2, while knowledge by acquaintance scored higher rankings of all the subsets in POP. This means the teachers gave more attention to the lived experiences of their learners and that the lessons were more experience-centred. The findings also reveal that lesson designs in lesson 1 improved, meaning that the teachers in the respective groups paid more attention to a phenomenological pedagogy and that the learners were more actively engaged in the teaching and learning process. Group 5 made the biggest shift in favour of a phenomenological approach, followed by group 4.

A Brief Description of the Respective Lessons and the Types of Knowledge Promoted

Group 1

Both lessons dealt with the periodic table. In lesson 1 the focus was on all the different elements on the periodic table. The lesson continued with the teachers explaining all the properties associated with the group numbers, differentiating between metals, non-metals, noble gases, solids, monatomic and diatomic elements, and ions. Given the theoretical nature of this lesson, mostly propositional knowledge was favoured. When the group was asked why they focused so much on abstract concepts and facts in their lesson, they stated:

> This method is new to us and we as teachers are more concerned with the content. After all, the learners must write examinations and we want them to be ready for it. But we will give it more thought and use it in our lessons.

Seeing that little attention was given to the lived world experiences of the learners or the use of thought-provoking scenarios, almost no link was made with knowledge by acquaintance.

In their second lesson on the same topic they focused more on how the objects that surround us are linked to the periodic table. They then proceeded by giving a brief overview of how the different elements react chemically to form various objects common to the learners. For example, they used things that the learners are acquainted with, such as water, carbon dioxide, oxygen and so forth. They then continued by explaining how each of these chemical substances is necessary to survive. When asked in the post-lesson discussion how they felt about the lesson, they said they are beginning to enjoy it, because 'we no longer have to be so textbook bound and prepare our lessons using the internet, YouTube videos, and so forth. It's amazing how much we learn ourselves'.

Group 2

The first lesson focused on the atoms and the subatomic particles surrounding the atom. At first they introduced the lesson by drawing a model of the atom with all its constituents—protons, electrons, neutrons and so forth. They then proceeded by explaining how these small subatomic particles can be used to produce an electrical current. As in the case of group 1, the lesson was mainly focused on propositional knowledge, but occasionally links were made with the real world that learners could connect with. When the group was asked why they did not concentrate more on the usefulness of the atom in their everyday experiences, they said:

> That is not so easy, but we will try to consult more sources next time. In fact, it is also hard work. This method is not who were are (all members shaking their heads), I know what works in the classroom and how my learners learn, but I feel like someone else when I teach this way. For now I first want to understand this method as it is new to me.

In their second lesson they paid more attention to the uses of atoms. They played a video on how atoms can be used in almost every aspect of our human existence. They invited a discussion and probed the learners (teachers pretending to be learners) on whether they have ever seen, touched or smelled an atom. The lesson continued with a classroom discussion in which the learners could talk about the origin of the atom. One teacher in the group said that this method actually lets you talk less; he exclaimed, 'I don't feel so tired after the lesson.' Another one said, 'By using their experiences and connecting it with the content makes them think about their everyday actions. Is that not what education is about?'

Group 3

The first lesson was on acids and bases. The teachers brought different household substances such as vinegar, soap and oranges to the classroom and placed them on the table. At this point they invited a discussion on how these substances/objects relate to acids and bases. The teachers then explained why certain substances such as vinegar can be ingested as opposed to soap and so forth. This evolved into a discussion on pH scales and on how to measure the pH of the substances they brought to school. This lesson leaned more towards being learner- and lived-world-centred. In the process the teachers promoted knowledge by acquaintance, which they successfully linked to propositional knowledge. When they were asked how they felt about having to change their teaching style, they laughed and said: 'We must start somewhere, but we still have a lot to learn. This approach will also give us an opportunity to learn from them and their experiences.' The second lesson followed the same approach, but this time the teachers used different products.

Group 4

Lesson 1 focused on chemical bonding. The teachers used the different objects in the room to describe and explain the bonds that hold these substances together. They then proceeded by breaking an object into smaller pieces so that the learners could see the invisible parts that make up the objects. They explained that the energy required to break the substance apart is the same as the energy of the bonds that keep the various particles together. They also used table salt as an example to explain how the molecules are interconnected and bonded together to form a compound. The second part of the lesson followed the same trajectory, but this time they introduced the lesson by using ice blocks, water and a cup of boiling water so that the learners could relate what it is that keeps the molecules together. They then explained the meaning of all the different concepts such as ionisation energy, bond length, bond strength and so forth.

Group 5

The lesson was introduced by drawing an oxygen molecule on flipchart paper as follows:

The teachers then discussed the atomic (proton) number and atomic mass and their relation to the element. For example: 'Can you see the small number and the bigger number in the box? The small number is the atomic number and the big number is the mass number. The number eight means the number of protons found in the nucleus, and the difference between the number of protons and the atomic mass number gives us the number of neutrons.' They then drew three different oxygen elements on the flipchart with varying mass numbers and kept the atomic number constant, namely, and explained the difference in the number of neutrons, hence introducing the term 'isotopes'. They continued the lesson by using another element following the same explanation as they did with oxygen. This lesson only promoted propositional knowledge and no consideration was given to the knowledge by acquaintance. After the lesson the group, almost in one voice, said:

> We know this is not what phenomenology is about, but once we understand the content a little better we can focus on its applications, right. Because it is about understanding how it fits in the world outside. We will definitely need new textbooks to assist us here, because the ones we use do not explain these things.

In lesson 2 they started to think more actively and first explained where the different isotopes of oxygen are found in nature. The learners were given time in their groups to discuss whether they have encountered the different types (isotopes) of oxygen, such as ozone, solid oxygen and so forth. After that we gave the learners five minutes to exchange ideas and search for information on their cell phones and to report back. The class had to find out what the physical differences were—such as colour, smell, shape and so forth. This was a drastic improvement in their design and implementation by shifting the focus to knowledge by acquaintance. Learners (peers) were given enough time for discussion in their groups and between groups to share information. After the lesson they stated that they think they are starting to understand what is expected from them and starting to place more emphasis on what the learners know about the topics.

Summary of the Post-lesson Interviews

In the post-lesson feedback sessions the teachers were given enough time to critique each other's presentations. They were encouraged to point out any contradictions and/or misconceptions, and whether or not the presenters adopted a phenomenological pedagogical approach to teaching the various topics. In the post-lesson interviews group 2, in agreement with group 1, felt that two challenges will impede on their effective implementation of phenomenology: (i) the need to prepare for the examination; and (ii) a shortage of time. They were afraid that if they focus too much on the learners' lived world, they might not have enough time to complete their syllabus and that the learners might not be ready for the examination. After their second attempt group 3 acknowledged and thanked the facilitators for introducing them to phenomenology and their input in their presentations. They stated that this method will help them to give more careful consideration to their learners' lived world experiences and the type of knowledge they promote. In the second round of lessons the teachers expressed an appreciation of the learner-centred instructional types that embrace the real-life experiences of the learners. What was consistent, though, was that they firmly supported each other's lessons and provided constructive feedback to each other. For example, group 5 stated after their second lesson that the more they engage with the phenomenological method, the more they like it.

Discussion and Implications

This study investigated the effect of a professional development workshop on Physical Science teachers' pedagogical practices. More specifically, the study looked at whether teachers changed their teaching approaches from the traditional to the phenomenological approach. The teachers' lessons were evaluated against POP. The study also investigated the types of knowledge promoted in each lesson. In addition, data were also presented on the post-lesson discussions during which the teachers could give their views about why they did what they did in every lesson.

Prior to the first workshop offered to the teachers, the DBE raised concerns that the participants were stuck in traditional teacher-centred pedagogies and textbook-driven approaches. This was confirmed in the first workshop, when the teachers stated that they have given up trying to come up with different strategies, because the learners are just not interested in learning. A review of the literature shows that a number of studies provide evidence that teachers lacked teaching strategies that 'excite' and 'engage' science learners (Kazeni & Onwu, 2013). Other studies indicate a lack of pedagogical knowledge, confidence and the ability to develop an appreciation of science (Basson & Kriek, 2012; Dube & Lubben, 2011; Kisiel, 2013;). This two-year long PDP in collaboration with the DBE shows that the teachers improved in their understanding of phenomenology as a teaching methodology—as they could design more lived-experience-centred pedagogies that promote greater knowledge by acquaintance, as opposed to traditional pedagogies that promote propositional knowledge. The findings also revealed that when teachers pay more attention to knowledge by acquaintance, learners become actively engaged in the teaching and learning process, and learn to understand the scientific processes behind their everyday experiences with objects as well as connect with nature at a much deeper level (Dahlin, 2001; Winch, 2013).

The data based on the POP scores in conjunction with the post-lesson discussions showed that the teachers wanted to change their pedagogies. Both groups 4 and 5 stated that they enjoyed the new method of teaching and will definitely use it in their classrooms, while groups 2 and 3 stated that, although the method is time consuming and distracts them from preparing learners adequately for examinations, they will give it more thought and use it in their lessons. Hence the two main impediments to adopting the phenomenological method were the teachers' strong focus on teaching for the examination and, equally important, lack of time. Kazempour (2009) reported similar challenges in his PDPs with science teachers.

Recommendations

If we want teachers to change their traditional pedagogical approaches, a starting point must be to introduce them to alternative pedagogies through PDPs. The findings in this study have shown the impact that new pedagogies can have on teachers as they are introduced to different knowledge types and different ways of practising science in order to prepare their learners more effectively by engaging them actively. Encouraging teachers to try out new ways of delivering their content should be one of the main aims of any PDP. According to Klein (2001), PDPs are the best way to bring about reform in education at large.

Some of the teachers pointed out the various impediments such as time constraints and the need to prepare for examinations. This study has shown two reasons why teachers do not want to change their pedagogies and therefore become stuck in traditional pedagogies. Therefore this study suggests that further engagement with the teachers is warranted to determine why they find it difficult to change their practices. This information is important to designers of PDPs so that they can give consideration to the design of their programmes when offering workshops to teachers. Gee (2001) points out that teachers find change very difficult to adjust to, because at times it demands that they be someone they are not. As the teachers in group 2 stated, this method made them feel different: 'It is not who I am.' According to Gee (2001), teaching is consistent with identity and by asking teachers to change is to divorce the person from how he or she sees themselves.

Lastly, curriculum planners and developers need to identify the impact the curriculum could have on the teachers. For example, the knowledge types have a direct bearing not only on what to teach but also how to teach. What was also highlighted in this study was the type of knowledge required if we want to expose learners to deeper levels of understanding and doing science. For example, if teachers must focus on knowledge by acquaintance, it is imperative that they get to know their learners, so that they can draw from the lived world of the learners to design their lessons. This implies that teachers must change their pedagogies accordingly, otherwise the objective of curriculum change is defeated.

The small sample size of this study limits the generalisability of the results. Follow-up studies could use randomised, controlled designs to explore how a PDP can successfully change teachers' pedagogical practices. The results suggest that feedback is very important for a PDP. Professional developers should consider using social media as support mechanisms to further encourage change in the teachers' pedagogical practices. Through social media they can monitor and assist teachers in the effective use of phenomenology as a teaching approach. Phenomenology is a highly complex field, but constant support via social media such as Facebook and Twitter can help teachers to incorporate this approach into their daily practices as Physical Science teachers.

References

Aoki, T. (1992). Layered voices of teaching: The uncannily correct and the elusively true. In W. F. Pinar & R. Irwin (Eds.), *Curriculum in a new key: The collected works of Ted Aoki.* London: LEA Publishers.

Basson, I., & Kriek, J. (2012). Are grades 10–12 Physical Science teachers equipped to teach Physics? *Perspectives in Education, 30*(3), 110–122.

Bennett, J., & Holman, J. (2002). Context-based approaches to the teaching of Chemistry: What are they and what are their effects? In J. K. Gilbert, O. de Jong, R. Justi, D. Treagust, & J. H. Van Driel (Eds.), *Chemical education: Towards research-based practice* (pp. 165–184). Dordrecht: Kluwer.

Brandom, R. (2000). *Articulating reasons: An Introduction to Inferentialism.* Cambridge, MA: Harvard University Press.

Centre for Development and Enterprise [CDE]. (2010, September). *The maths and science performance of South Africa's public schools: Some lessons from the past decade.* CDE research no. 1. Johannesburg, South Africa.

Dahlin, B. (2001). The primacy of cognition—or of perception? A phenomenological critique of the theoretical bases of science education. *Science and Education, 10,* 453–475.

Dass, P. M. (2001). Implementation of instructional innovations in K-8 science classes: Perspectives of in-service teachers. *International Journal of Science Education, 23,* 969–984.

Department of Basic Education. (2003). *Revised national curriculum statement for Physical Science.* Pretoria: Government Printer Works.

Department of Basic Education. (2008). *National curriculum statement: FET Physical Science.* Pretoria: Government Printer Works.

Department of Basic Education. (2011a). *Curriculum and assessment policy statement.* Pretoria: Government Printer Works.

Department of Basic Education. (2011b). *Revised national curriculum statement.* Pretoria: Government Printer Works.

Department of Education. (1999). *Curriculum 2005.* Pretoria: Government Printer Works.

Department of Education. (n.d.). *National Assembly Training and Education Department (NATED) interim core syllabus for Physical Sciences (HG, SG & LG).* Pretoria: Government Printer Works.

Dube, T., & Lubben, F. (2011). Swazi teachers' views on the use of cultural knowledge for integrating education for sustainable development into science teaching. *African Journal of Research in Mathematics, Science and Technology Education, 15*(3), 68–83.

Foundation for Research and Development. (1993). *South African science and technology indicators.* Pretoria: Author.

Gee, J. (2001). Identity as an analytic lens for research in education. *Review of Research in Education, 25,* 99–125.

Halling, S. (2002). Teaching phenomenology through highlighting experience. *Indo-Pacific Journal of Phenomenology, Special Issue, 3,* 1–6.

Heidegger, M. (1967). *Being and time* (J. MacQuarrie & E. Robinson, Trans.). London: SCM Press.

Heidegger, M. (2002). *The essence of truth* (J. Sadler, Trans.). London: British Library of the Congress.

Kazempour, M. (2009). Impact of inquiry-based PD on core conceptions and teaching practices: A case study. *Science Education, 18*(2), 56–68.

Kazeni, M., & Onwu, G. (2013). Comparative effectiveness of context-based and traditional approaches in teaching genetics: Student views and achievement. *African Journal of Research in Mathematics, Science and Technology Education, 17*, 50–62.

Kisiel, J. (2013). Introducing future teachers to science beyond the classroom. *Journal of Science Teacher Education, 24*(1), 67–91.

Klein, B. S. (2001). Guidelines for effective elementary science teacher in-service education. *Journal of Elementary Science Education, 13*(2), 29–40.

Koopman, O. (2017) *Science education and curriculum in South Africa*. New York, NY: Palgrave Macmillan.

Koopman, O. (2013). *Teachers' experiences of implementing the Further Education and Training science curriculum*. Unpublished doctoral dissertation. Stellenbosch University, Stellenbosch.

Koopman, O., Le Grange, L., & De Mink, K. (2016). A narration of a physical science teacher's experience of implementing a new curriculum. *Education as Change, 20*(1), 89–111.

Lee, C., & Krapfl, L. (2002). Teaching as you would have them teach: An effective elementary science teacher preparation program. *Journal of Science Teacher Education, 13*, 247–265.

Locke, J. (2009). *Of the abuse of words*. London: Penguin Books.

Naidoo, P., & Lewin, K. M. (1998). Policy and planning of Physical Science education in South Africa: Myths and realities. *Journal of Research in Science Teaching, 35*(7), 729–744.

Østergaard, E., Dahlin, B., & Hugo, A. (n.d.). From phenomenon to concept: Designing phenomenological science education. *6th IOSTE Symposium for central and Eastern Europe*

Piburn, M., Sawada, D., Falconer, K., Turley, J., Benford, R., & Bloom, I. (2000). *Reformed Teaching Observation Protocol* (Tech. Rep No IN00-1). Tempe, AZ: Arizona State University, Arizona Collaborative for Excellence in the Preparation of Teachers.

Pinar, W. F., & Reynolds, W. M. (1992). Curriculum as text. In W. F. Pinar & W. M. Reynolds (Eds.), *Understanding curriculum as phenomenological and deconstructed text* (pp. 1–13). New York, NY and London: Teachers College Press, Columbia University.

Reddy, V. (2006). The state of mathematics and science education: Schools are not equal. In S. Buhlungu (Ed.), *State of the nation: South Africa, 2005–2006* (pp. 392–416). Pretoria: HSRC Press.

Shulman, L. S. (1986). Those who understand: Knowledge growth in teaching. *Educational Researcher, 15*(2), 4–14.

Shulman, L. S. (1987). Knowledge and teaching: Foundations of the new reform. *Harvard Educational Review, 57*(1), 1–22.

Shulman, L. (2005). *The signature pedagogies of the professions of law, medicine, engineering and the clergy: Potential lessons for the education of teachers*. Unpublished paper delivered at the Math Science Partnership workshop in Irvin, California.

Tyler, R. (1949). *Basic principles of curriculum and instruction*. Chicago, IL: University of Chicago Press.

Van Manen, M. (1990). *Researching lived experience: Human Science for an action sensitive pedagogy*. New York, NY: State University of New York Press.

Winch, C. (2013). Curriculum design and epistemic ascent. *Journal of Philosophy of Education, 47*(1), 128–145.

CHAPTER SEVEN

Towards a Humanising Philosophy of Education in South Africa

Introduction

On 14 October 2016 the television channel SABC2 paid homage to Thuli Madonsela after her seven-year tenure of the position of Public Protector (PP) of South Africa came to an end. The office of the PP is a Chapter 9 institution, whose role is to act as a watchdog over the Constitution of the country (for full details, see Chapter 9 institutions in the Constitution of South Africa). One of the duties of the Public Protector is to ensure that public officials act in accordance with the law and discharge their duties properly. During her term in office she was often praised for her courage to tackle challenges such as the misappropriation of state funds, especially by high-profile state officials, including the President, Jacob Zuma. On 15 October 2016 eNCA news broadcast live the annual Ahmed Kathrada Foundation lecture at the University of Johannesburg, at which occasion Madonsela elaborated on the current crisis plaguing the country, including issues such as (i) social inequality, (ii) '#what must rise for fees to fall', (iii) corruption of government officials and the consequences of this, amongst other things. Two salient points in her keynote address inspired the writing of this article, that is, (i) the values of *ubuntu* must rise, and (ii) the need for humane values to drive our existence in South Africa.

Her speech reminded me of Julia Annas (1993), who writes 'in ancient ethics the fundamental question is, How ought I to live? Or what should my life be like?' In short what is our ultimate goal in life? She then draws from Aristotle to answer the question and states that each of us should attempt to live a good life. The least we could do, Aristotle maintained, is 'to act well' (Ibid., p. 27). Such a view seems plausible when we observe that it does not preclude acting on the basis of ordinary reasons of self-interest. In Plato's *Republic* Socrates identifies the underlying principles upon which an ideal society should be built. He asserts that the first principle is 'mutual need', because man is not self-sufficient and therefore unable to supply all his external needs without the assistance of another. The second is based on differences in aptitude between men. Socrates concludes that different people are good at different things. Consequently, each man must develop the ability to master that which he is good at. To do so, Socrates points out, is to invoke the full possession of praxis (moral good of the other) and peiosis (knowledge, skills and method). This raises the question: 'How do Africans view a good and moral life?'

To Africans humanity is described by what Le Roux (cited in Le Grange, 2012, p. 331) terms: 'umzimba (body, form, flesh); umoya (breath, air, life); umphefumela (shadow, spirit, soul); amandla (vitality, strength, energy); inhliziyo (heart, centre of emotion); umqondo (head, brain, intellect) and ubuntu (humanness)'. The humanness referred to in ubuntu finds expression in a communal context rather than in individualism. There is a sea of literature on the African ethic of *ubuntu* (for a full account, see Higgs, 2012; Gyekwa, 1997; Le Grange, 2012; Letseka, 2000; Sindane, 1994; Waghid, 2004). Le Grange (2012, p. 331) explains that ubuntu refers to brotherhood (translated: 'I am because of others'). He avers that ubuntu has a 'normative connotation embodying how we [humans] ought to relate to each other in Africa'. For centuries Africans has been raised and trained in the moral obligation that they have towards other Africans and the environment. This moral obligation is his or her belief that they have to do 'good' for the 'good' of others and they have to think of themselves as being bound up with others. As Sindane (1994) writes: 'Ubuntu inspires us to expose ourselves to others, to encounter the difference of their humanness so as to inform and enrich our own' (pp. 8–9). Letseka (2000) alludes to the IsiZulu expression of *umuntu ngumuntu ngabantu* [translated literally as: a person is a person because of others]. Here Letseka avers that in African ways of life, a person depends on others just as much as others depend on them. This points to the interdependence of humanity on the African continent. Consequently, Africans that strive to fully embrace ubuntu are driven by a humanist concern to treat others with a sense of fairness and care. This means such a person's everyday actions is motivated by an act of care, thoughtfulness underpinned by a state of awareness not to harm the other.

Similarly, the Sotho concept of *ukama*, which is a broader concept than *ubuntu*, refers to the relatedness of the African to nature and the cosmos (Le Grange, 2012). *Ukama* represents a form of closeness and affection that Africans have with the world within which they find themselves. Bujo (1998, p. 22) writes: 'The African is convinced that all things in the cosmos are interconnected.' Knowledge that was framed within the tradition of *ubuntu* and *ukama* has always been a spiritual activity. This spiritual directedness of knowledge via *ubuntu* and *ukama* signifies the virtues of self-control and treating others with respect. Furthermore, it underscores strong beliefs that require a presence of the self (self-awareness) and the other, respect (for others and the environment), empathy for the other, valuing relationships as well as spiritual and emotional well-being. All these aspects, implicit in the concepts of *ubuntu* and *ukama* point to a particular mindset/philosophy that formed the basis of African education systems whether formally (in tribes) or informally (in families/communities). But how does all of this resonate with education in South Africa?

The education of the traditional Africans was, firstly, not based on autonomy or individualism, but rather on values such as respect and cordiality. Respect for oneself and the other were at the heart of moral conduct. Second, their primordial understanding of education was grounded in 'child-rearing'—that is, bringing up children to be able to take up their position in their respective societies. Such education is contradictory to the modern view of education as formal and systemic, focused on strong academic development and conceptual abilities with little care and concern for humane values and the environment. If *ubuntu* is the root of African philosophy as a way of being (Ramose, 2004), one may wonder what the value of education is, in the absence of care and respect for others and the environment. For modern Africans, *ubuntu, botho* and *ukama* have been redefined as role-specific duties that a parent, teacher, fire-fighter, policeman and so forth has to fulfil. The new meaning and standards afforded to ubuntu and ukama related to citizenship represent a new form of being that Peters (2005) refers to as 'government citizens'. This so-called 'government' citizen is described as 'self-regarding', one who 'acts through a series of instrumental, temporary bilateral relationships' (Peters, 2005, p. 136). In this context very little, if any, attention is paid to what it means to be human or a morally good citizen as one of the main objectives in society. Hodgson (2010) describes this type of education as the search for the economic life, which Davids and Waghid (2016) argue is designed to prepare the learner and the student for a market-driven work culture. This neoliberal agenda within a capitalist system forms the cornerstone of modern man's educational training. Hence, success is measured in the accumulation of wealth at the cost of values such *ubuntu* and *ukama*. This raises the question, which also brings me to the aim of the article, of whether we can escape such a system, and if the answer is yes we can, then what can we replace it with to bring back the values of *ubuntu* and *ukama* to South Africans.

This article reports on an investigation into the question: Can a humanising philosophy of education in South Africa overcome the complex web of neoliberal governmentality? This study therefore critically examines the significance of neoliberal governmentality for teaching and learning. I start by drawing on Michel Foucault's concepts of *governmentality* and *biopower* to show how complex organisations such as schools and universities manipulate the way Africans think and view the world. From this perspective I engage philosophically with government's neoliberal agenda of 'high knowledge' and 'high skills' and how this promotes the consumerist interests of corporate power through commodified knowledge, resulting in the dialectical negation of both learners and students. Through the elaboration of a humanising philosophy, drawing from Husserl, Heidegger and Merleau-Ponty, I attempt to show that teachers should re-think their focus on knowledge as an end and encourage new ways of thinking.

The Rise of Neoliberal Governmentality in South Africa

Christie (2006) explains that when the African National Congress (ANC) became unbanned in 1990, various discourses of education policy were formulated to articulate the possibilities and limitations of educational change in South Africa. She averse that the National Education Policy Investigation (NEPI) used the format of policy options to explore all the different possibilities of what an education system, that embraces the values of democracy, should look like. This policy was maintained in the Implementation Plans for Education and Training, coordinated by the ANC's Education Desk and the Centre for Educational Policy Development. In the mid-1990s Christie (2006) notes, these policies came under scrutiny when education theorists and researchers began to question what had happened to the envisaged policy after the establishment of the new Government of National Unity (GNU) in 1994. The vision to shift to a 'People's Education' agenda of the 1980s was vehemently critiqued as it did not constitute a coherent set of policies as it largely focussed on the classroom. de Clercq (1997) pointed out that the proposed policy by the GNU was flawed in its conceptualisation of policy processes and cautioned that the new policies might not meet the ANC's objective of redress. She writes: 'these policies are in danger of creating conditions that will assist the privileged education sector to consolidate its advantages while making it difficult for the disadvantaged to address their problematic political realities' (de Clercq,

1997, p. 127). Since then different policy themes have been explored by academics such as Motala and Pampallis (2001), Sayed and Jansen (2001), Chisholm, Motala, and Vally (2003) through policy analysis of the policy processes and policy shifts as well as systemic changes in education in post-apartheid South Africa.

The final outcome in the analysis of educational policy processes was a shift in the vision the ANC had for education in the 1980s and 1990s for the people of South Africa. Instead of following through with their vision as a 'Liberation Movement' for a 'People's Education' they adopted a shift from liberation to government. Commensurate with this shift to 'government' was the adoption of a framework that promoted procedures, regulations, and domains of knowledge. In changing its status from a banned 'terrorist' organisation to an elected government the ANC wanted to show that they can think and act like a modern state. According to Christie (2006) the main objectives of the ANC were to build governmental capacity through issues of population, economy, and security by engaging with particular technologies of practice and domains of knowledge. With this decision they shifted their attention to the macro-economic arena in order to build confidence in the economy and to establish legitimacy and capacity as a leader on the African continent and a player in world affairs. The shift to government meant the adoption of practices and procedures that favours neoliberal governmentality at the expense of core ethical African values such as *ubuntu, batho and ukama*.

The official demise of the repressive apartheid system in 1994 gave birth to the Department of Education's Curriculum 2005 (C2005) (Department of Education (DoE), 1997) and all the Department of Basic Education's (DoBE) other curricula that followed in its wake. These included, for example, the Revised National Curriculum 2005 (RNCS) (DoBE, 2002), the National Curriculum Statement (NCS) (DBE, 2006), and the Curriculum and Assessment Policy Statement (CAPS) (DBE, 2010), all of which were regarded as the educational route out of the sterility of apartheid education. Central to all these curricula was the belief that the learner and his or her lived world should be placed at the centre of the teaching and learning process (DoE, n.d., 1997; DoBE, 2003, 2006, 2010). Whether this materialised at the classroom inter-phase is another concern.

Several South African studies over the last two decades show that despite the wonderful principles on which these curricula are built teachers frequently encourage rote learning (Koopman, 2013; Le Grange, 2016; Ngamu, 1991). In such a paradigm, learners rely heavily on memorisation to pass examinations. According to Chisholm and Wildeman (2013, p. 90), this approach to teaching comes as no surprise since high-stakes examinations already existed as early as 1915. The authors further point out that, after the transition to democracy in 1994, South Africa adopted a 'post-burcaucratic model' of accountability in which the quality of teaching was measured by learner performance as a strategic plan to monitor teachers. The emphasis placed on learner performance as a measure of teacher effectiveness encourages teachers to neglect the foundational knowledge learners require to master any subject (Chisholm & Wildeman, 2013). This, they argue, undermines the purpose of education. Such a system leaves learners underprepared with no confidence or competence in Science. This target-setting agenda encourages teachers to 'teach to the test'. In so doing, teachers confuse the learners by providing obscure and limited knowledge of the respective disciplines. Other studies that investigated the lived experiences of teachers found that teachers struggle to effectively teach their subjects as they are governed by strict time frames and the need for good pass averages (Koopman, 2013).

Le Grange (2016) points out that the content in subjects such as humanities; mathematics and science are derived from countries from Europe and or the global north, despite the growing body of knowledge generated over the last 22 years by African and South African theorists. Although the student demographics have changed significantly at our universities the curriculum makers have not changed thereby the status qua remains. The focus on high knowledge and high skills dominate curricula with little consideration given to the lived world of learners and university student.

The Complex Web of Neoliberal Governmentality in Education

'Governmentality' presents a genealogy of the question of government and explains how governments establish successful social control over their citizens (Foucault, 1991, p. 91). In this essay Foucault alludes to the task of government in its role of establishing continuity in both an upwards and downwards direction in the ruling of its citizens. Foucault (1991) writes: 'This downward line, which transmits to individual behaviour and the running of the state, is just at this time beginning to be called police...' (p. 91). Thus, Foucault is clearly making the connection between the relationship of self and governing authorities. Applying this Foucauldian notion of 'governmentality', 'policing' and 'control of the subject' can be seen in the role of the state with its prescriptive curricula, district offices that govern the principal in ensuring that state designed curricula are implemented, the principal in relation to the teacher and the teacher as implementer of the curriculum and his or her duty in delivering the content to the learners. In other words governmentality is not only concerned with the work of governing others but also with the work of the self, which is how teachers conduct themselves. Governmentality dates back to medieval times (Foucault, 1980). During this period Western societies delegated all power to the ruler, whose duty it was to decide what was legal (right or wrong) for societies to do. When the Roman Empire was constructed, the emperor had a constitutive role to play as he had absolute power to determine the laws and rules of the empire. In European societies not under Roman rule, the King was the 'central personage in the entire legal edifice of the West' (p. 94). In modern societies, absolute rule or domination over society has devolved from one absolute ruler to groups such as governments, that is, absolute power has been decentralised and become shared power. Foucault states that governments are focused on:

> a sort of complex composed of men and things…men in their relations, their links, their imbrications with those things that are wealth, resources, means of subsistence, the territory with its specific qualities, climate, irrigation, fertility, and so on; men in their relation to those other things that are customs, habits, ways of acting and thinking, and so on; and finally men in their relations to those still other things that might be accidents and misfortunes such as famine, epidemics, death and so on. (cited in Hodgson, 2010, p. 112)

Christie (2006) explains how the modern-day government of South Africa are not very different and have similar intentions for their people. The governments, she points out, have put in place particular organisations, practices, rationalities and doctrines through which they govern, leading their populations to meticulous forms of 'governmentalities'. The term 'government' or 'governmentality' ('gouverner') in a broad sense refers to specific techniques and procedures that direct and shape human behaviour (Christie, 2006). They are imposed on people by governments through their organisations, practices, and complex sets of knowledge domains. According to Peters (2005), to Foucault the word 'governmentality' means 'mentality of rule' as a means to signify the emergence of a distinctive mentality of rule that he alleged became the basis for modern liberal politics.

Peters (2005) states that modern-day governments use neoliberal technologies in organisations such as schools and universities through education, science, law and so forth to exercise control over their citizens and manipulate them. School and university curricula are constantly policed through increased monitoring of learner/student performance through increased systemic evaluations, international benchmark tests, e-portfolios, and so forth (Koopman, Le Grange, & De Mink, 2016). This notion of surveillance and control does not stop at school or university level. It is now extended deep into adulthood under reform agendas such as lifelong-learning espoused in curricular documents (see National Curriculum Statement, 2008; Curriculum and Assessment Policy Statement, 2011) and the charter of graduate attributes of universities. This means that, in the bigger scheme of things, multinational businesses and global markets that drive educational programmes in South Africa have a wider pool of human capital. In other words, 'human' has now been replaced with 'human capital', which in the neoliberal setting the value of a human being is measured in terms of monetary value.

In such a system institutions such as schools and universities become engines of economic reform within the ethos of capitalist competition. The upshot of this hierarchy of institutional control is a deeply divided society that does not want social equality, but rather a belief in the equalising power of profit-driven neoliberalism. This equalising power of neoliberalism is predicated on the assumption that cognition and information generation should be the most important function of schools and universities. For example, at school level learners are prepared for careers in science through government-designed curricula. In this hegemonic neoliberal system the learners and students are viewed as consumers of 'commodified' knowledge and in the process become 'technicians of learning' (Davids & Waghid, 2016). These government-designed curricula with their predesigned accounts of knowledge do not promote the kind of creativity and open-mindedness that encourage learners and students to think differently, but they rather use them as objects, to borrow from Foucault, through which 'habit 'inhabits' the body as a form of capillary power that permeates the body and inserts itself in action, attitude, discourses and everyday lives' (Foucault, 1982, p. 127). This means that schools and universities promote hierarchies and an acceptance of social relations of domination and subordination, which implies that neoliberalism cannot promote equalisation. In addition, schools and universities are expected to respond to employer needs. Bowles and Gintis (1976) argue that this mythological view of education devalues the pillars necessary for developing an educated citizenry. This leads to the next question: How does neoliberal governmentality shape the way people think? To explain this, I will turn to Foucault's (1978) notion of human 'biopower'.

Biopower: The Emergence of the 'Technologised Self'

In a neoliberal system *desire* and *self interest* as an economic agenda become the driving force of the technologisation of the 'mind' of South African learners. This can be seen as the emergence of a new kind of power which Foucault (1978) refers to as '*biopower*'. Biopower is a system in which the human body (the learner, in this context) is viewed as a central component in the operation of power relations. Biopower does not replace governmentality—government's rule, control and manipulation over its citizens—but rather overlaps with it. According to Foucault (1978), biopower is a normative force/power that rules over life itself. Its target is human bodies that aim to imbue certain subjectivities with a certain form of 'naturalness'. Naturalness is a shift in the notion of rationality of interest to the individual as a self-centred being that limits the power of the state over the human being. This, Foucault explains, is skilfully done by developing and encouraging 'desire' for any human want as the emergence of new form of power (biopower) through a new science of human behaviour emanating from the political economy. In a lecture in 1978 Foucault explains how biopower shapes 'desire' as a way of becoming as follows:

> We could also say that the naturalness of the population appears in a second way in the fact that this population is of course made up of individuals who are quite different from each other and whose behavior, within a certain limit at least, cannot be accurately predicted. Nevertheless, according to the first theorists of population in the eighteenth century, there is at least one invariant that means that the population taken as a whole has one and only one mainspring of action. This is *desire*. (Foucault, 1978, p. 89)

Foucault alludes to the powerful influence of *biopower* and the strong hold of the *political economy*. He describes how 'self-interest' and 'desire' can be used as an epistemic discourse that drives human behaviour in society. These constructs can generate powerful emotions that not only infiltrate the social, political and economic practices of society but also the moral fibre of every individual. For example, the manipulation of desire is a neoliberal mechanism that has infiltrated schools and universities that in the process promotes from a perspective that intensifies the subjective intrusion of neoliberal governmentality. To this end learners and students are enticingly taught and encouraged to act out of desire and are consequently socialised to become powerless against the strong emotional forces of desire. In other words, schools and universities do not prepare learners and students to become emancipated or to deal with knowledge intelligently and critically such as expressing care for the other or the environment. For example, when teaching a topic such as nuclear physics, teachers and lecturers do not embed the subject matter in its wider social, environmental and ethical context. Their lessons do not include issues around the proliferation of nuclear weapons, nuclear safety, or a discussion around the environmental impact of nuclear energy on the environment. Learners should be made aware of the crippling effect that nuclear weapons and energy have, not only on the environment but also on human freedom and movement. For example, learners should be made aware how Einstein spoke out on nuclear strikes against Japan and how he in hindsight regretted having urged an atomic weapons program in the United States. Instead, these issues are ignored while learners and students are given a one-dimensional view of the technical aspects of the knowledge as opposed to the multidimensional value of the knowledge. When this happens, students do not see their discipline within a wider context to become more aware of how science impacts on society.

Odora-Hoppers and Richards (2011) remind us that, on the African continent, the first generation of colonialism involved conquering the physical spaces and bodies of Africans. By contrast, the second generation of colonialism involved colonising the African mind through disciplines such as education, science, economics and law. Colonisation of the mind takes place when externalised impulses of the economic life created by the neoliberal agenda penetrate the mind and succeed in their sheer power to defeat the self (Wa'Thiongo, 1986), leading to a 'technologisation of the self'. This Wa'Thiongo points out is not always easy to spot in our educational experiences, because the brute force of neoliberalism can be described as a seductive system, To be effectively seduced means the person is persuaded or convinced to conform to a particular mindset and of course, submit to desire. In other words seduction is accomplished through the participation of the one who is seduced, much like an illicit love affair that is driven by the inclinations of desire. Thus conceived, learners find themselves caught up in a powerful complex web of seduction that aims to 'technologise the self' by silencing the inner voices of reason (around), forcing them to surrender to a lifestyle of self-interest and desire. This means our way of life and view of the world become distorted and redefined for the so-called 'modern' good in society. But what is this modern 'good' worth if we lose our identity and subjective connectedness with the world in which we live. Here Wa'Thiongo (1986) reminds us of the importance of taking a step back and search for understanding of who we are and why we are, we will always be trapped in the cold and confusing view of the world not knowing what imperialism and the modern day form of imperialism called 'neoliberalism' has done to us. Next I discuss how we can overcome this powerful web that engulfs our subjectivities.

Overcoming the Powerful Web of Neoliberal Governmentality

Human Consciousness as an Inclusive Philosophy of Lived Experience

Philosophers such as Husserl, Heidegger and Merleau-Ponty spent their life's work underscoring the preeminent significance of human consciousness for humanity. Husserl's (1970) book entitled *Ideas pertaining to pure phenomenology*, Heidegger's (1967) magnum opus *Being and time* and Merleau-Ponty's (1962) *Phenomenology of Perception* have one common theme, that is, without human consciousness there is no world. While Husserl (1970) launched the school of philosophy devoted to the systemic analysis of objective consciousness, Heidegger (1967) points out that the world cannot be inferred but is what every human being starts with. Merleau-Ponty's (1962) work has shown that the world presents itself at each and every instance as a meaningful totality in which every moment possesses part of a larger dynamic whole. Consciousness, according to Merleau-Ponty (*ibid.*) consists of the intentional totality of our human existence. Thus consciousness is not simply *our* world but *the* world. This implies that conscious is nothing but embodied experiences. Embodied experiences can be described as the 'felt space' or 'body in world and world in body', and without them the world is an imaginary place of which the person has no sense of awareness. Therefore anybody of knowledge bereft of experience is at best described as mythical. This is because lived experience is a pre-condition for learning and language is used to conceptualise and articulate every moment. Each moment or event is expressed in language in which the words receive their precise sense from the experience. In other words, language as a whole forms part of the discourse, but is not the discourse in its totality. Searle (1995) adds a different stance to our existence and points out that without people or any other sort of sentient beings, language would not exist. Searle elaborates on this point by explaining that concepts like a hammer, a tree, a car and so forth would not exist if there had never been human beings. What humans attempt to do through experience is to articulate with precise descriptions through language the intrinsic properties of objects relative to the intentionality of agents. Therefore discourse is embedded in experience that takes place in a social setting in the world. This brings Merleau-Ponty to the realisation that it is only through experience that a person can discover the world which possesses him or her more than the experience itself.

The above implies that the nature of experience philosophically has to be the mediating principle in which learning must be rooted. Experience in the teaching and learning process forms the ultimate poles of subject and object. From this perspective Langan (1966, p. 22) writes:

> The practical orientation of commonsense experience towards clear consciousness tends to polarise our attention, noetically, towards the most willful active principle—the subjective Ego as doer, grasped as intellect and will—and, noematically, toward the most structured object—the finished product of our praxis-centred daily concerns, the intellectually fixed, objectivised thing. While both the voluntary ego and objective thing implicitly depend on the actual body's living in the world and making an experience possible in the first place, the bodily synthesis nevertheless goes about its task so silently, so fundamentally, that its transcendental contribution is no more noticed than the light which illumines and thus makes possible every visible spectacle.

In other words, apart from using our experiences to describe the world, when lived experience becomes the philosophical approach to teaching and learning, it has the potential to transport individual (learner/student) beyond the narrow bounds of the technical and method towards an illuminating new realm of existence. In this realm the learner transcends his or her old ways of knowing and doing to discover new worlds of ideas and thinking in which a person sees the world afresh from different perspectives. In other words, subjects like mathematics, science and history should not be presented or seen as bodies of knowledge to be learned that lead to mythical understandings or virtual realities, but as coming to grips with the world with its content. This will allow teachers to start a subject like mathematics, science or history with question about the learners' perceptions about of the content, as oppose to teaching as the delivery of ready-made knowledge. The fact that individuals have different experiences in the world allows them to bring different perspectives to learning subjects like mathematics and science. These different experiences and perspectives brings with them opportunities for new dialogues about the world and the usefulness of subjects like mathematics, science or history. When a learner or student can apply the knowledge that emerges from these subjects, he or she can look at events in the world in new ways to understand each moment through the looking glass of experience. When this happens, experiences of the same events are no longer routinely or consistently viewed as if the world controls the body involuntary, but the individual learns to take more control of his or her world. Thus, when the teaching of mathematics and science is rooted in experience, it allows the individual to discover the world as a 'world-founding principle'. This way of learning makes a phenomenon under discussion or analysis more transparent, thereby revealing more than one way to resolve an inquiry.

Heidegger's (1967) students described his teaching style as nothing short of 'electrifying' (Kruger-Ross, 2015). He was renowned for his commanding presence both as a lecturer and public speaker. Instead of handing down 'commodified' knowledge or predesigned knowledge, he would guide his students into the being of the phenomenon in question. His students, including Hans-Georg Gadamer, Karl Loöwith, Walter Biemel and Hannah Arendt, describe his teaching approach as an ontological dissection of phenomena. He would start every lecture by laying out his thesis and taking the title of the announced lecture or course, sometimes word-by-word, and completely transform the common sense meaning as a way to reveal new insights into thinking about 'being' or thinking and 'being' (Kruger-Ross, 2015; Magrini, 2014). In this way the lecture became not a description of the work, but a source or a guideline for subjective inquiry. This approach did not only inspire his students to move away from grasping old ideas, cold facts and descriptions of life, but encouraged them to take a step back and to reflect on their experiences and to carefully apply the main thesis to their subjective lived worlds. Heidegger's humanising philosophical approach inspired his students to search for truth, making the discovery of new ideas possible. In other words his students had to use the information taught and to go beyond what they have learned. This approach helped them to break free from the confinements of the known to discover a new world, new ways of thinking and looking at life. It also assisted them to create and understand a common humanity. Heidegger's (1967) approach to teaching or public speaking as sketched by Kruger-Ross (2015) resonates well with his notion of 'dasein'—'being there' or 'there being'—because to him the essence of our 'being' is existence or living (1967, p. 207). By asking questions instead of teaching teachers relate to the changing temporal structure that occurs through various stages that learners go through in life.

Vandenberg (2008) notes that after Heidegger published 'Being and Time' in 1927 he became immediately famous. This encouraged many students from all over Europe came to study with him. After departing from Heidegger and his teachings they formed 'a first, second and third generation of existentialist phenomenologists' (p. 253). This is because the existentialist asks the question of being. Concerning Heidegger's view of technology Vandenberg points out that the only way to see things in their original being in modern society is through art also referred to as the aesthetic perception. In all other facets of life technology has taken over. This raises the question: How do we learn when we make experience or being the starting point and end point of our lessons/lecturers? I will now draw on the work of Deleuze and Merleau-Ponty to explain this.

How Do We Learn in Experience?

Deleuze (1994) explains how we learn when experience is used as a primer for new information. He points out that an individual does not carry into action some *a priori* representation of the world as away to understand phenomena. Deleuze sees learning as a matter of indexing all possible conjunctions relating to an experience of an event. For example, in swimming the body's spatial orientation and subconscious mind are subjected to a confined space. In the event of swimming the body is viewed in conjunction with a wave: a body and wave. Deleuze writes:

> When a body combines some of its own distinctive points with those of a wave, it espouses the principle of a repetition which is no longer that of the *same* but involves the *other*—involves difference, from one wave and one gesture to another, and carries that difference through the repetitive space thereby constituted. To learn is indeed to constitute this space of an encounter with signs, in which distinctive points renew themselves in each other, and repetition takes shape while disguising itself. (Deleuze, 1994, p. 23)

In other words, it is not only our mind but the multiple and varying parameters of the body that continues to create novel relations in our real experiences. Learning happens when a body actualises its virtual potencies, leading to new understandings (Deleuze, 1994). It is only through real engagement with waves where the virtual 'essences' or the 'concept' of swimming exists, thereby allowing the individual to understand the notion of swimming. Experience is therefore paramount for learning in order to create new ideas and therefore attention needs to be paid to places and spaces in our teaching. Drawing from Merleau-Ponty's (1962) lived body theory, we do not only perceive a real life with our eyes, our minds or any other bodily organ—we perceive things with the sum of the parts of the body. The being [the self] and being [in the world] form the two poles of Merleau-Ponty's thought (Rabil, 1967). Indeed, Rabil (1967, p. vii) concurs that 'neither the subjectivity of the self nor the objectivity of the world, but the relationship between them…continuously renders the self more than subjective and the world less than objective…'. The mind therefore becomes subordinate to the body as the senses help the body search for a new understanding on the basis of previous events and experiences of a particular phenomenon. This brings me to the value of the senses in learning.

The Value of Sense Experience

Each person born into this world experiences a plurality of phenomena. In search of meaning and understanding, a person uses his or her human faculties such as his or her sensory and mental apparatus. Merleau-Ponty's (1962) analysis and understanding is explained *in fine* detail how phenomena are perceived in the world through the senses. He points out that in search of meaning people decipher events to search for consistencies and repetitions in their world. For example, consistencies refer to properties of objects such colour, shape, size and various others. Other consistencies include gestures, movements, expressions and so forth. These properties can only be found in the comportment of the body in the world and the body's connectedness to objects in the world. The search for consistencies is important for the human mental faculties because people use them to construct routines or generic protocols to give structure to their lives. Consistency, also referred to as 'constancy', makes the body intentional, since it is always on the outlook for a consistent world to harmonise an experience as perceived by the senses with every other moment it has experienced. Merleau-Ponty (1962) avers that this search for consistency explains Husserl's (1975) *Urdoxa* ('first' or 'primary' doctrine) of intentionality. To Merleau-Ponty the prime effect of 'intentionality' should be viewed from the perspective of the body's place in the world. He holds that the 'intentional body' manifests itself as a spontaneous presumptuous unity of consistency. This raises the following question: Does every person experience the world of phenomena in the same way?

Conversely, what a person experiences in the world, is not necessarily what is out there in the world, but depends on the nature of a person's sensory organs. For example, Koopman (2017) states that when tasting red wine a person depends on his or her sensory organs of sight, smell and taste. He then breaks down the role of the various sense organs and explains how sight is used to identify the colour and texture of the wine, smell to identify the tannins and bouquet of the wine, and taste to decipher all the ingredients such as its oral sensations and fruity flavours. However, oral sensations are complex and depend on various factors such as ethnicity, age, gender and overall health in addition to other physiological factors (Pickering, Moyes, Bajec, & Decourville, 2010). Physiological factors, such as sensitivity to 6-n-propylthiouracil (PROP) and bitterness, cause some individuals to perceive the taste and tactile sensations of red wine more intensely than others. These factors make the description of wine based on oral sensations even more challenging. In other words, the nature of our experiences, such as our sensory faculties, are subject dependent. Thus reality is divided into two parts:

1. There are things as they are in themselves, which is independent from their being experienced. We have no means of accessing such things.
2. There are things as they appear to us in the world of appearances. That is as things present themselves to us through experience. Things can also be is presented as a science which represents our total world.

In other words, our senses play an important role in our perception of the world and hence information. For this reason MacMurray (2012) states that acquiring high knowledge and high skills in order to be adequately prepared for the world of work is not a holistic education. In his view, this is not even the most important task of education; as he puts it: 'It is rather the minimum that an industrial society must demand for efficiency's sake' (p. 773). This is because this type of education does not add value to an individual's humanity; instead it marginalises the individual from being conscious of the self. He suggests a type of learning that harnesses the full use of the senses, but in a contemplative and disciplined way (MacMurray, 2012). He contends that to be contemplative about sense experiences involves leading a creative life as opposed to a dull life of intellect dominated by concepts and definitions. He claims that applying the senses is another way of learning to be human, which stimulates creativity. It is a child's birthright to be creative, so as to allow the imaginative abilities to stimulate the child's intellectual faculties. To be creative and imaginative requires a deeper connectedness to the senses, which requires a person to see, hear, feel, smell and taste. These sense experience activities root the individual in the present, allowing him or her to tap into a deep connectedness with his or her surroundings. In other words, knowledge becomes a way of life that flows from the outside to the inside, allowing the body to become an extension of the mind (Merleau-Ponty, 1962). When this happens, he argues, the person can unlock the intrinsic worthwhile state of the individual. This helps the person to escape the predilection of neoliberal governmentality and move into the realm of life that gives him or her, a deeper sense of purpose.

Conclusion

In this chapter I attempted to evoke the significance of lived experience as an important starting point in the teaching and learning process. For the authors, human consciousness carries an individual's totality of experiences with the world. Therefore lived experience should form the fundamental basis of teaching as a way to illuminate a person's active-passive participation in the world. I have argued that to ask our learners and students to learn and think in a disembodied context creates no interest in dialogue. But when body and world are in dialogue, we learn to see and understand how each adjusts to the other under the impulsion of the situation. Such a dialogue is what education should be about as opposed to the neoliberal predilection for acquiring concepts and facts. The so-called 'high knowledge and high skills' approach, one of the seven principles that form the basis of South Africa's Curriculum and Assessment Policy Statement, cannot give texture and structure to the body and how to learn to connect with the mind. Instead, it brings greater separation between body and mind, which I have argued forms the basis of neoliberal governmentality. Such a paradox serves the interests of multinational business and the corporate world with a strong focus on consumerist market-driven knowledge.

In conclusion, I want to draw from the work of Hannah Arendt, who was a student of Martin Heidegger. To Hannah Arendt (1998) successful people are not what they have produced, but who they are. In her view true artists becomes their work and sees themselves as a mirror image of their work. She often draws from the world of the craftsman to illustrate all activities involved in creating the final product. Consider the carpenter and his activities. First there is the vision, that is, the mental model of the product that he works towards. Secondly, there are the various activities (work) in producing the product, for example a table or a chair. Most of these activities are directed towards finishing the product. But then there are times when he has to buy materials, clean his working area—for example, when he has to pick up the sawdust, clean and/or sharpen his tools and clean out his truck. During all these menial activities the carpenter's actions seem to be counterproductive. One can say that these days are his off-days; however, this is all necessary in realising the vision (product). Just as the carpenter takes a day or more off to do some clean-up work, teachers at times need to take their focus away from the planned curricular activities and administrative duties, and turn their attention towards the child as a living body. How the (child's) body relates to the space he or she occupies. Personalising the child's perspective through reflection of the consciousness made explicit. When this happens the teacher learns to see the child with a deep sense of sensitivity and thoughtfulness. To see another as a living being should not be viewed as a sacrifice but as part of his or her duties in the teaching and learning process. Just as the carpenter at times admires his product by admiring it or feeling the smoothness of the surface, and so develops an internal appreciation for his own work. In the same way teachers need to embrace the experiences of the learner. Simply focusing on knowledge as an end in itself would be living an impoverished life as a teacher. It is perhaps unfortunate that in South Africa good teachers are identified on the basis of good results in tests and examinations, thus driving the neoliberal agenda of high knowledge and high skills. In order to achieve this goal successfully, the teacher may not see the learners as persons, but as useful bodies that serve the agenda of the state.

References

Annas, J. (1993). *The morality of happiness*. New York, NY: Oxford University Press.

Arendt, H. (1998). *The human condition* (2nd ed.). Chicago, IL: University of Chicago Press.

Bowles, S., & Gintis, H. (1976). *Schooling in capitalist America: Educational reform and the contradictions of the economic life*. New York, NY: Basic Books.

Bujo, B. (1998). *The ethical dimension of community*. Nairobi: Paulines Publication.

Chisholm, L., Motala, S., & Vally, S. (2003). *South African education policy review, 1993–2000*. Johannesburg: Heinemann.

Chisholm, L., & Wildeman, R. (2013). The politics of testing in South Africa. *Curriculum Studies, 45*(1), 89–100.

Christie, P. (2006). Changing regimes: Governmentality and education policy in post-apartheid South Africa. *International Journal of Educational Development, 26*, 373–381.

de Clercq, F. (1997). Policy intervention and power shift: An evaluation of South Africa's education restructuring policies. *Journal of Education, 12*(3), 127–146.

Davids, N., & Waghied, Y. (2016). *Educational leadership in becoming: On the potential of leadership in action*. London: Routledge.

Deleuze, G. (1994). *Difference and repetition* (P. Patton, Trans.). New York, NY: Columbia University Press.

Department of Basic Education. (2002). *Revised national curriculum statement: Grades R–9 (Schools): Natural Sciences.* Pretoria: Author.

Department of Basic Education. (2003). *Physical Sciences national curriculum statement. Grades 10–12 (General policy).* Pretoria: Author.

Department of Basic Education. (2006). *Physical Sciences national curriculum statement: Grades 10–12 (General).* Pretoria: Author.

Department of Basic Education. (2010). *Curriculum and assessment policy statement: Physical Sciences (CAPS).* Pretoria: Author.

Department of Education. (n.d.). *National Assembly Training and Education Department (NATED) interim core syllabus for Physical Sciences (HG, SG & LG).* Pretoria: Author.

Department of Education. (1997). *Curriculum 2005.* Retrieved from http://www.polity.org.za/govdocs/misc/curr2005html

Foucault, M. (1978). Governmentality' (Lecture at the Collège de France, Feb. 1, 1978) (pp. 87–104). In: G. Burchell, C. Gordon, & P. Miller (Eds.), *The Foucault effect: Studies in governmentality.* Hemel Hempstead: Harvester Wheatsheaf.

Foucault, M. (1980). Power/knowledge. In C. Gordon (Ed.), *Selected interview and other writings 1972–1977.* Brighton: Harvester Press.

Foucault, M. (1982). The subject and power. In H. Dreyfus and P. Rabinow (Eds), *Michel Foucault: Beyond structuralism and hermeneutics.* Chicago IL: University of Chicago Press.

Foucault, M. (1991). Governmentality. In P. Rabinow (Ed.), *The Foucault reader: An introduction to Foucault's thought* (pp. 1–22). Harmondsworth: Penguin.

Gyekye, K. (1997). *Tradition and modernity: Philosophical reflections on the African experience.* Oxford: Oxford University Press.

Hodgson, N. (2010). Narrative and social justice from the perspective of governmentality. *Journal of Philosophy of Education, 43*(4), 559–572.

Heidegger, M. (1967). *Being and time* (J. Macquarrie & E. Robinson, Trans.). London: SCM Press.

Higgs, P. (2012). African philosophy and the decolonisation of education in South Africa: Some critical reflections. *Educational Philosophy and Theory, 44*(2), 37–56.

Husserl, E. (1970). *The crisis of the European sciences and transcendental phenomenology: An introduction to phenomenological philosophy* (D. Carr, Trans.). Evanston, IL: North-Western University Press.

Husserl, E. (1975). *The Paris Lectures* (P. Koestenbaum, Trans.). The Hague: Martinus Nijhof.

Koopman, O. (2013). *Teachers' experiences of implementing the FET Science curriculum.* Unpublished PhD Thesis. Stellenbosch University.

Koopman, O. (2017). Science education and curriculum in South Africa. New York: Palgrave Mcmillan

Koopman, O., Le Grange, L., & De Mink, K. (2016). A narration of a Physical Science teachers experiences in implementing a new curriculum. *Education as Change, 20*(1), 149–171.

Kruger-Ross, M. (2015). Raising the question of being in education by way of Heidegger's phenomenological ontology. *Indo-Pacific Journal of Phenomenology, 15*(2), 1–12.

Langan, T. (1966). *Merleau-Ponty's critique of reason*. New Haven and London: Yale University Press.

Le Grange, L. L. (2012). Unbuntu, ukama, environment and moral education. *Journal of Moral Education, 41*(3), 329–341.

Le Grange, L. L. (2016). *Decolonising involves more than simply turning back the clock*. The Conversation Africa Pilot.

Letseka, M. (2000) African philosophy and educational discourse. In P. Higgs, N. C. G. Vakalisa, T. V. Mda, & N. T. Assie-Lumumba (Eds.), *African Voices in Education*. Cape Town: Juta.

MacMurray, J. (2012). Learning to be human. *Oxford Review of Education, 38*(6), 661–674.

Magrini, J. M. (2014). *Social efficiency and instrumentalism in education: Critical essays in ontology, phenomenology and philosophical hermeneutics*. New York, NY: Routledge.

Merleau-Ponty, M. (1962). *Phenomenology of perception*. London: Routledge.

Motala, E., & Pampallis, J. (2001). *Education and equity: The impact of state policies on South African education*. Johannesburg: Heinemann.

Nganu, M. (1991). *Overview of African countries' strategies in tackling problems of science, technology and mathematics education in human resource development for the post-apartheid South Africa*. London: Commonwealth Secretariat.

Odora-Hoppers, C. A., & Richards, H. (2011). *Rethinking thinking*. Pretoria: Unisa Press.

Peters, M. (2005). The new prudentialism in education: Actuarial rationality and the entrepreneurial self. *Educational Theory, 55*(2), 267–282.

Pickering, G. J., Moyes, A., Bajec, M. R., & Decourville, N. (2010). Thermal taster status associates with oral sensations elicited by wine. *Australian Journal of Grape and Wine Research, Issue 16*, 361–367.

Sayed, Y., & Jansen, J. (Eds.). (2001). *Implementing education policies: The South African experience*. Cape Town: University of Cape Town Press.

Searle, J. (1995). *The construction of social reality*. New York, NY: Basic Books.

Sindane, J. (1994). *Ubuntu and nation building*. Pretoria: Ubuntu School of Philosophy.

Rabil, A. (1967). *Merleau-Ponty: Existentialist of the social world*. New York, NY: Columbia University Press.

Ramose, M. B. (2004). In search of an African philosophy of education. *South African Journal of Higher Education, 18*(3), 138–160.

Vandenberg, D. (2008). A guide to educational philosophizing after Heidegger. *Educational Philosophy and Theory, 40*(2), 249–267.

Waghid, Y. (2004). African philosophy of education: Implications for teaching and learning, *South African Journal of Higher Education, 189*(3), 56–64.

Wa'Thiongo, N. (1986). *Decolonising of the mind: The politics of language in African literature*. Portsmouth, NH: Heinemann Publishers.

Index

OMPLICATED

A BOOK SERIES OF CURRICULUM STUDIES

Reframing the curricular challenge educators face after a decade of school deform, the books published in Peter Lang's Complicated Conversation Series testify to the ethical demands of our time, our place, our profession. What does it mean for us to teach now, in an era structured by political polarization, economic destabilization, and the prospect of climate catastrophe? Each of the books in the Complicated Conversation Series provides provocative paths, theoretical and practical, to a very different future. In this resounding series of scholarly and pedagogical interventions into the nightmare that is the present, we hear once again the sound of silence breaking, supporting us to rearticulate our pedagogical convictions in this time of terrorism, reframing curriculum as committed to the complicated conversation that is intercultural communication, self-understanding, and global justice.

The series editor is

Dr. William F. Pinar
Department of Curriculum Studies
2125 Main Mall
Faculty of Education
University of British Columbia
Vancouver, British Columbia V6T 1Z4
CANADA

To order other books in this series, please contact our Customer Service Department:

(800) 770-LANG (within the U.S.)
(212) 647-7706 (outside the U.S.)
(212) 647-7707 FAX

Or browse online by series:

www.peterlang.com